SCIENCE AND CREATION
IN THE MIDDLE AGES

Science and Creation in the Middle Ages

Henry of Langenstein (d. 1397) on Genesis

NICHOLAS H. STENECK

UNIVERSITY OF NOTRE DAME PRESS

NOTRE DAME LONDON

Library of Congress Cataloging in Publication Data

Steneck, Nicholas H
Science and creation in the Middle Ages.

Bibliography: p.
1. Science, Medieval. 2. Science–History.
3. Heinrich von Langenstein, 1325 (ca.)-1397. I. Title.
Q125.S7425 509'.023 75-19881
ISBN 0-268-01672-0

Manufactured in the United States of America

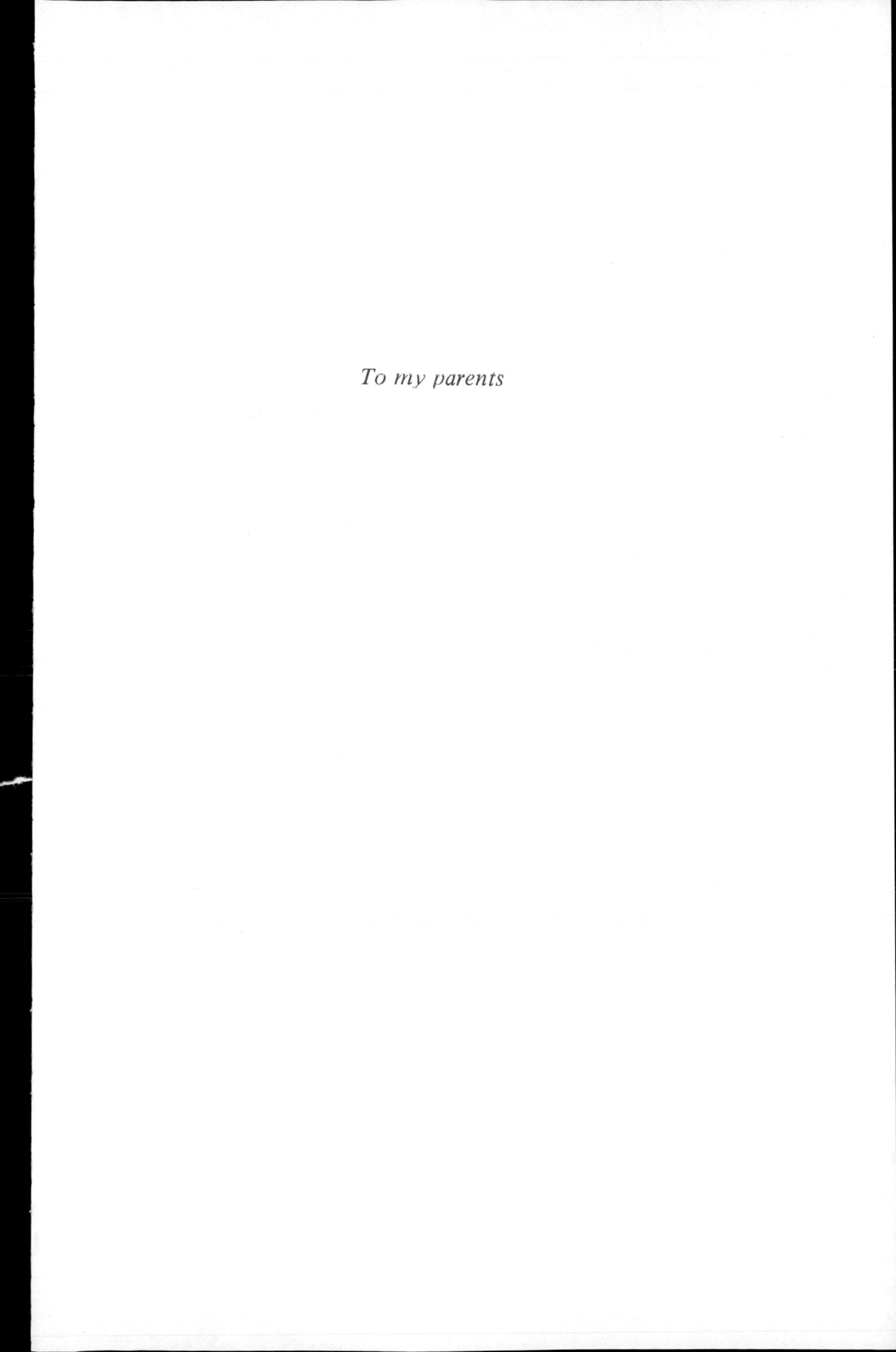

To my parents

Contents

Illustrations

Acknowledgments

The effort it has taken to complete this study has been aided greatly by a number of colleagues and granting institutions. Many of the sources and manuscripts listed in the bibliography were drawn together by the diligent efforts of two research assistants, Thomas Burns and David Lux. Grants from the Horace Rackham School of Graduate Studies at the University of Michigan, the National Science Foundation, and the American Council of Learned Societies provided the time and funds to pursue my research and see it through to publication. Marvin Becker, William Courtenay, David Lindberg, and Charles Trinkaus read the manuscript in preparation and provided many helpful suggestions. Charles Trinkaus, in particular, has taken a constant interest in my work, and I am deeply grateful for his help and encouragement. Dr. Astrik Gabriel of the Medieval Institute at Notre Dame, whose own research on Henry's life and writings provided both inspiration and guidance, kindly placed at my disposal his own microfilm collection of Langenstein manuscripts and the excellent facilities of the Institute. Finally, a special note of thanks is due to my colleagues at Michigan whose firm conviction that historians should be able to communicate with one another, even historians who deal with the obscure recesses of medieval science, prompted me to present this study in such a way that its object and content could be understood by more than a limited number of specialists. During my years at Michigan I have profited greatly from the lively atmosphere that this conviction inspires.

Sent.	*Quaestiones quarti Sententiarum,* ms. Alençon, Bibliothèque de Ville 144, fols. 1ra–140va.
Sermo	*Sermo de Sancta Katharina Virgine,* ed. Albert Lang, 1948.
CSEL	*Corpus Scriptorum Ecclesiasticorum* Latinorum, 85 Vols., Prague, 1866–.
PG	*Patrologiae Cursus Completus,* Series Graeca, ed. J. P. Migne, 166 Vols., Paris, 1857–.
PL	*Patrologiae Cursus Completus,* Series Latina, ed. J. P. Migne, 221 Vols., Paris, 1844–.

Introduction

Sometime around the middle of the fourteenth century a master of arts at Paris by the name of John Buridan (d. after 1358) made the less-than-profound observation that "if anyone is moved in a ship and he imagines that he is at rest, then, should he see another ship which is truly at rest, it will appear to him that the other ship is moved."[1] The obvious point that Buridan was attempting to make is that change can only be measured in reference to some fixed or known background; without a reference point, events that seem to encompass change may in fact be illusions that stem from uncertain points of observation. This generalization, which had such a profound impact on so many areas of thought when it was set forth in a different form by Albert Einstein over five hundred years later, has important ramifications for historical analysis. Historians are no more immune to deceptive relative judgments than are physicists. Without fixed reference points, historical change cannot be measured.[2]

A case in point, indirectly related to this study, concerns the changes that epitomize the emergence of modern science. Historians generally assume that the ship containing such important personages as Copernicus, Vesalius, Kepler, Harvey, Galileo, and Newton is the one that is the most intimately connected with the birth of modern science. Through its course of action, this vessel brings to port the events of the Scientific Revolution. However, few historians have endeavored to determine the overall course of the other ship, the one containing the science of the late Middle Ages and Renaissance. The resulting hiatus in scholarship is crucial. If Buridan's generalization has bearing on the study of history—and I believe that it does—it means that we have insufficient knowledge about the background to the Scientific Revolution to make judgments concerning its dynamics. Since this last assumption may not be readily apparent, some explanation is in order.

The background against which the changes of the Scientific Revolution must be measured is extensive and extremely complex. It includes at least three hundred years of university history that witness the slow turn of the open intellectual atmosphere of the medieval university into the commonly criticized atmosphere of its sixteenth-century counterpart. It includes the emergence of extra-university institutions and the many philosophies that became popular in these institutions as a result of the fifteenth-century

Renaissance. It includes the spread of literacy and culture down through the ranks of society to a rapidly rising artisan class and the subsequent fusion of an artisan mentality–in the form of an emphasis on improvement, progress, and empirical utilitarianism–with the activities of "high culture." It includes the widening of intellectual horizons through the discovery of new lands and new ways of looking at the past. Without a recognizable past and a sense of change over that past, hope for an improved future, the Baconian ideal, could not have been realized. These and a host of additional factors that bring in religion, philosophy, sociology, and so on, make up the context against which the "modernity" of minds such as Copernicus's or Galileo's must be measured. They provide the points needed to fix precisely the crucial changes that occurred over the course of the fourteenth, fifteenth, and sixteenth centuries in the formation of the modern scientific mind. Though it may seem trivial to suggest that such points need to be thoroughly studied before judgments can be made about the origins of modern science, it is unfortunately the case that they have not been studied thoroughly enough to permit sound judgments to be made.

The latter generalization holds true especially for the period of the history of science that comprises the immediate focus of this study, late medieval science. While few historians would take issue with the assumption that medieval and modern science are different, neither those who deal mainly with the Scientific Revolution nor those whose primary focus is medieval science have endeavored to present descriptions of this important background in ways that allow comparisons to be at all comprehensive. The reason for this, in my opinion, is that neither group has attempted a sufficiently broad and methodologically unrestricted analysis of late medieval science. Instead, the science of this period has, for the most part, been selectively picked over by both groups in an effort to uncover what are assumed to be crucial continuities and/or discontinuities between it and modern science. Not enough attention has been given to the broad specturn of works and ideas that *in toto* comprise the science of the late Middle Ages and to the overall cultural context within which these works and ideas reside.

Historians of the Scientific Revolution, including many historians whose primary interests lie broadly in Renaissance culture, have chosen either to ignore medieval science, under the assumption that it is largely unrelated to the development of modern science, or to construct brief and at times inadequate syntheses of the complex traditions involved for the purpose of setting later developments out in sharper contrast. The reasoning that lies behind the former approach is set out quite plainly by A. Ruppert Hall in his introduction to the "Rise of Modern Science Series." To quote Hall:

> In a history of modern science it is unnecessary to describe the slow and devious process by which, after the fall of the Roman Empire, Greek

> science (with some accretions) was partially recovered and assimilated in Europe. On the other hand, it is very important to analyse the effect that the fresh exploration of Greek sources had on the fifteenth and sixteenth centuries, when mediaeval science seemed to have become sterile.[3]

For Hall and other historians of science, the forces needed to account for the rise of modern science are not found in medieval science. Therefore, there is no need to pursue what is not relevant. The elements of change, according to this methodology, can be set out without comparative studies.

The problem with this approach to the history of early modern science is that very few scholars find it possible to write about the Scientific Revolution without making at least some comparative judgments. And when they do so, especially if these judgments are offhand and not central to the main argument, they run the risk of setting forth misstatements and misinterpretations. As careful a historian as Charles Gillispie, who admits that most "modern writers," including himself, have not "combined in their own understanding the sympathy, the scholasticism, the Latinity, . . . and the gothicism which would permit them to penetrate the true spirit of Copernicus's life-work," is still able to conclude that "the revolutionary inspiration" that lay behind sixteenth- and seventeenth-century developments in cosmology and physics "came from the Platonic criticism playing a ray of mathematical realism upon the vast mass of verbal distinctions into which Aristotelian natural philosophy had proliferated in the late Middle Ages and from which it endlessly drew theological tidbits out of nature."[4] It goes without saying that even more knowledge of "Latinity and scholasticism" would be needed to prove the closing statement than to reach an understanding of Copernicus's life-work. And yet Gillispie is not alone in including such brief, undocumented digressions in his analysis for the purpose of setting the Scientific Revolution more clearly apart from what came before.

Far more serious than these passing comments are the straightforward and often inadequate attempts of historians of Renaissance science and the Scientific Revolution to meet the issue head-on and set later scientific thinkers sharply apart from the scholastics. In doing so, all too many assume the bias of their sources—Petrarch and others clearly thought that their own ideas and methods differed from those of the scholastics[5]—and thus dispense with late medieval science on very questionable grounds. Eugenio Garin in his informative study of *Science and Civic Life in the Italian Renaissance* explains what is new in Renaissance thought vis-à-vis the medieval world view without once quoting a scholastic or limiting his comparisons to a specific segment of medieval science.[6] E. A. Burtt in his influential *Metaphysical Foundations of Modern Science* supports a questionable interpretation of scholastic thought with a long passage from Dante and then uses this as the foundation for a discussion of the modernity of Copernicus, Galileo, and Newton.[7] Hugh Kearney in his recent study, *Science and Change*, bases his characterization of medieval science on the false premise that it "stressed the role" of "final cause" and on a misleading analogy to "organic" development.[8] And so it goes. Whole cases are frequently built on straw men who either never existed or who represent only a fraction of the complex tradition that makes up the science of the late Middle Ages. As a result, the comparative judgments based on these straw men tie the assumed changes of the Scientific Revolution to "ghost ships" of questionable authenticity. The ghostly nature of these ships becomes even more elusive as the issues involved are expanded to include questions of mechanism, man's place in the universe, the evolution of the experimental method, and so on.

For their part, historians of medieval science have not done as much as would be hoped for to aid other historians in studying the Scientific Revolution. Since its inception over seventy years ago under the guiding hand of Pierre Duhem, the study of the science of the late Middle Ages has had as its major objective the investigation of those aspects of medieval science that have ties with the most salient features of modern science.[9] Since the identity of these salient features is not universally agreed on, the range of investigation within medieval science has admittedly been broad, extending from studies of particular sciences and physical theories to investigations of method and select aspects of the medieval world view. However, it has not been broad enough, nor has it succeeded in developing methods that bring to light the full range of generalizations needed to understand the Scientific Revolution.

First of all, there exist no really comprehensive analyses of the scientific thought of even the most important late medieval scientists. Surveys of the science of such thirteenth-century thinkers as Aquinas and Albert the Great have not been matched by similar studies of any major fourteenth-century scientists. Even when a relatively large number of edited or printed sources is available for a single figure, as is the case with the writings of John Buridan and Nicole Oresme (d. 1382), no comprehensive assessment of their thought has been attempted. As a result, whereas it is essential for historians of later periods to have ready access through secondary literature to general statements about medieval science (they cannot be expected to pursue extensive primary research on the Middle Ages and still remain historians of later periods), such general statements have yet to be formulated. The background C. S. Lewis provides for historians of literature in his *Discarded Image*, or Brian Stock for historians of twelfth-century science in his study of Bernard Silvester, has not been duplicated for the science of the late Middle Ages.[10]

Moreover, the forward-looking orientation of the history of medieval science has left vast segments of this science unexplored and many crucial questions unanswered. Very little attention has been devoted to understanding why medieval science failed; that is, why it did not realize the potential of some of its ideas and methods, and what prompted the many fruitless discussions (by modern standards) that fill the folio pages of Aristotelian commentaries. We know in specific instances and with regard to individual problems what motivated the medieval scientist, but we do not know where in general his interests lay. Were they in nature, in debating, in the study of God's creation, in teaching, or in some other aspect of science? Such issues must be resolved with special reference to the entire corpus of medieval science. It is as important to understand how and why a particular figure sought to account for the trivial and insignificant (by our standards) as to understand how he explained gravity and projectile motion. Both come under the domain of natural philosophy, and so both are part of the late medieval world view.

Finally, insufficient attention has been given to the actual practice of medieval science as opposed to philosophical pronouncements about how science ought to be conducted, and to working world views as opposed to metaphysical assumptions that have potential bearing on world views. The methodological and epistemological suggestions of William of Ockham (d. 1347) and Buridan have been examined with regard to their potential revolutionary impact on science, but there has been very little effort to demonstrate that the methods by which scholastics approached the study of nature changed significantly over the course of the fourteenth century. Are such developments simply literary or philosophical suggestions that have little connection to the actual practice of science? This seems to be the case with Ockham's theory of intuitive cognition, which failed to dislodge the standard doctrine of species from psychological treatises. Or did they in fact change the way science was done? Many possible feathers remain to be put into the caps of fourteenth-century scientists; but until these scientists can be shown to have broken free from tradition in their conduct of science and their world views, such feathers must remain only possibilities.

Given these deficiencies, the primary objective of the following study is to take a cautious first step toward articulating a historical method that will broaden our understanding of late medieval science and help place it more certainly within the general development of early modern science. The essence of this method is neither revolutionary nor complex. My intent is to set out, in as readable a form as possible and keeping in mind the needs and interests of both specialists and nonspecialists, a general survey of late medieval science as it appears in the writings of one very important late medieval scientist. In doing so, I have attempted to remain sensitive throughout to the categories and priorities that the system of science itself demands. My chapters and divisions are drawn, for the most part, from the organization of a major document, Henry of Langenstein's *Lecturae super Genesim.* When I deviate from this organization, I explain the reasons for doing so. At the same time, I have worked under the supposition that the best way to find out what medieval science is all about is to observe it in action on as many fronts as possible. I am not convinced that what the medieval scientist, or any other scientist, said about the practice of science has any necessary connection with the procedures that are used in its pursuance. Nor am I convinced that the philosophies of particular ages have any necessary connection with the everyday and working world views of these ages. In both cases, it is the latter that I am after, the actual procedures used in pursuing science and the working world view that underlies these procedures. I have searched for these within the broadest historical context possible. By doing so, I hope to provide an example of the sort of analysis that seems to me essential for understanding the history of, and background to, the Scientific Revolution.

In addition to this broad objective, this study has a more limited, but at the same time more obvious, goal. The selection of Henry of Langenstein as the focus for this book was not without reason. Henry's scientific writings, and particularly his *Lecturae* on the opening chapters of Genesis, are unusually comprehensive for the period. The great age of medieval encyclopedias had fairly well come to a close by 1300, before the interesting developments in science of the fourteenth century. Thus, Henry's writings provide an encyclopedic corpus to study for an age in which encyclopedic works were not common. Moreover, this particular scientific corpus has yet to receive very much attention. No systematic analysis of the *Lecturae* has even been attempted. In fact, this important source, which has circulated in numerous manuscript copies, has never been edited or printed and is rarely used in secondary works on Henry's thought or on any other aspect of late medieval thought. Apart from the neglect of his lectures on Genesis, little attention has been given to Henry as a scientific thinker. A few of his scientific writings have been the object of limited studies, but nothing has been written regarding his science as a whole. Were Henry a minor figure whose thought counted for little, this lack of scholarship might be understandable. He was not, however, and therefore an important chapter in late medieval history has yet to be written. Thus, in addition to presenting a general analysis of late medieval science as viewed in its broadest context, this study also endeavors to bring to the attention of the scholarly world a survey of the scientific thought of Henry of Langenstein.

Of course, certain difficulties are encountered in surveying a single person's thought and then attempting to project out from this "representative" figure to the larger macrocosm of late medieval science. The gravest of these difficulties concerns the relevance of the ideas in the microcosm to the larger realm of the macrocosm. Most of the information that Henry relates in the *Lecturae* is grounded in discussions set out in his other scientific writings and in even broader histories of ideas that extend from the Greeks through Roman, Arabic, and later Latin sources to the high and late Middle Ages. I have attempted to bring in some of this broader context by noting, and at times discussing, portions of Henry's other scientific writings and of the writings of other figures that have bearing on his thought. Obviously, neither of these efforts has been exhaustive. It is impossible in a study of this nature to cover adequately the content of the *Lecturae* alone, much less the remainder of Henry's scientific and science-related works. Moreover, much of the history relating to the topics discussed in this book lies buried in manuscript sources that have not been studied. Only a few of these topics have been the object of careful scholarly analyses. There remains, therefore, a great deal more work to be done.

A second difficulty encountered in writing this book has to do with style. For the most part I am summarizing ideas and condensing arguments that are spread over the folio pages of the *Lecturae* and Henry's other scientific works. This has made it virtually impossible to include in the notes the critical apparatus that would normally be found in a study of this type. To include certain corroborative quotations would have required pages of Latin text to support ideas that are summarized in a single paragraph. Rather than doing this, and hence rendering this book out of the hands of all but a few specialists, I have attempted to provide some of the crucial terminology, where appropriate, in the text and to be specific in indicating where in the manuscript materials more extended discussions can be found. This will pose an inconvenience, but not an insurmountable obstacle, for scholars interested in pursuing some of my arguments in greater detail. Most of Henry's writings have circulated in multiple manuscript copies, and these are available in a number of major European libraries and in microfilm collections in the United States.

With this background and the aforementioned qualifications in mind, it can once again be noted that this book is intended primarily as a study of the other ship, the one whose crew does not bring science directly to port. In presenting this study, I have not limited my research and judgments strictly to the late Middle Ages. Like most historians of science, I have from time to time felt compelled, for better or for worse, to make assessments of Henry's thought relative to what was to follow. I have done so because the Scientific Revolution takes on different dimensions when it is viewed from the decks of this other ship. There are many things about our modern scientific world view that would have surprised Henry. At the same time, there are other things that would not. In sorting through these two categories of thoughts, I have not always come to believe that what we see to be the essence of the

Scientific Revolution would have been the things that would have most astounded Henry. He could have, I think, come to accept quite easily many of the principles of seventeenth-century mechanical philosophy. He might not have been all that shocked to find that atomism could eventually prove to be a fruitful way of explaining nature. However, other things about modern science would have been totally beyond his ken, and it is these unexpected aspects of science that I would point to as the crucial elements of the Scientific Revolution. More than any other factor, a firm belief that we are still far from defining these unexpected aspects explains why I have endeavored to become conversant with the thought of one late medieval scientist.

CHAPTER I

The Life, Training, and Works of a Medieval Scientist

If one can trust the information recorded at his death in 1397, Henry of Langenstein, also of Hesse, was born in 1325 in or around Langenstein in Hesse, Germany. Apart from this projected date, no information has as yet surfaced regarding the first half of his life. The earliest direct biographical information that is extant dates from early in 1363, when his name was recorded in the records of the University of Paris as a student in the arts preparing for his final degrees, which followed shortly thereafter. By July 18 of the same year he had apparently met all of the requirements and became a regent master in the arts.[1]

The next nineteen years of Henry's life (1363–1382) were spent at Paris pursuing what was a normal academic career for the day. Immediately after he received his master's degree he almost certainly went on to lecture in the arts. Throughout the 1360's and early 1370's his name was inscribed in the records of the University as a regent master in the arts sponsoring students for degrees. At the same time he wrote the majority, if not all, of his works devoted exclusively to science. Most probably it was also at some point in the 1360's that he began his theological studies—studies completed in 1376 when he was formally accepted as a regent master of theology. Hereafter his rise to a place of importance within the university community, and more importantly within the Church, was rapid. Two years after he received his master's degree in theology he was selected by the University as one of three spokesmen to go to Rome to present the University's position regarding the recent division of power within the Church, the Great Schism (1378–1415). At this time he became one of the most influential supporters of the Conciliar movement and a leading figure within the Church. Once drawn into the public forum in 1378, he would never again completely leave it.

Henry's years at the University of Paris came to an end in 1382 when he and a number of his German colleagues were forced by the political turmoil of the Schism to leave Paris and seek more hospitable lands. His subsequent journeys led first to a brief stay at the Cistercian abbey at Eberbach, where his friend and former colleague at Paris, James of Eltville, was abbot. Here, while "living between the oak and beech trees with the monks of the cloister at Eberbach,"[2] he apparently pondered the problems of the Church and theology, becoming, in the process, the proponent of the mystical theology that is elaborated in his famous and extremely popular *Speculum animae*. At the same time he paid his debt of gratitude to the brothers of the abbey by rereading the *Sentence* commentary composed by their abbot, James of Eltville, at Paris over a decade before. The peace and seclusion of the abbey did not keep him out of touch with other developments, particularly those relating to the Schism. Several important letters date from this period, including ones to Deacon Eberhard of Ippelbrunn, John of Eberstein, and to another close friend, Eckhard of Ders, the Bishop of Worms. He apparently visited Eckhard in 1383 when he left Eberbach for what was to become his final residence, Vienna.

The trip to Vienna was made at the invitation of Duke Albert III of Austria (1363–1395), who was seeking help in reorganizing and revitalizing the then inactive University of Vienna. Following his arrival in Vienna in 1384, Henry played an important role in this activity. He was undoubtedly one of the moving forces behind the new statutes of 1385 and assumed a variety of official duties within the University over the next twelve years of his life. In the year 1385 he also began his monumental *Lecturae super Genesim* in his official capacity as lecturer in theology. At the same time he never seems to have lost interest in the problems of the Schism, keeping up an active correspondence with numerous church leaders around Europe. If writings are any indication, his pace slowed very little, if at all, toward the end of his life. A flood of treatises, sermons, letters, and miscellaneous writings continued to flow from his pen throughout the 1390's and even up to his death in 1397.

Outlining the details of how and where science fits into this *curriculum vitae* presents a number of problems. Even though Henry was apparently older than most students when he began his arts training at Paris (the normal age for beginning arts studies would be the late teens), the details surrounding his life, falling into a very typical pattern for the period, remain obscure until well after his arts training–the medieval equivalent to a science training.[3] Thus, we know very little about his early career as a scientist or the order in which his scientific works were written.

In sections one and two of this chapter, these dual aspects of Henry's science are discussed in more detail in an effort to set the background for the *Lecturae super Genesim.* The chapter concludes with some remarks about the historical consciousness that Henry exhibits in his discussion of science in the *Lecturae.* The relevance of this last section may not at first be apparent. However, if it is recalled that medieval science is as much a historical activity as it is either a rational or experimental one, the need for such a discussion should be obvious. The laboratory in which Henry and most other medieval scientists worked was the lecture hall; the experimental apparatus they used was the book. Surveying his historical perspective thus becomes the process of describing the environment in which Henry, over the course of his lifetime, carried out his science.

1. Science Training and Career

The lack of information relating to Henry's early life leaves only one segment of his science training open to historical speculation, that is, his studies in the arts at Paris in and prior to 1363. Since the course of study required of arts students at this time is well documented, inferences about Henry's own course of study can be drawn by projecting from the general program to his specific case. This procedure, which is commonly used to establish the date at which important persons came to Paris, is not without difficulties. The statutes of the University present a somewhat ambiguous picture of the course of study in the arts. Moreover, statute requirements do not necessarily correspond to the actual courses of study that students followed. Nevertheless, some possible conclusions about Henry's early training can be drawn by taking a careful look at the arts program as it was organized and functioned in his day.

Promotions in the arts at Paris customarily came at three stages in a student's career: determination, licensing, and inception. (Determination came at the end of the years of study on subjects relating to the trivium; licensing was the formal process by which students who had gone on to audit lectures on the quadrivium and related subjects were granted the license to teach in the arts; inception broadly describes the events through which newly licensed masters were initiated into the ranks of teaching masters.)[4] At each stage, or with the granting of each degree, specific requirements pertaining to lectures audited, years in residence, age, and so forth had to be met.[5]

In theory, then, if we know when a student received a particular degree, it is possible to reconstruct from the requirements the course of study that "must have been followed" prior to the granting of that degree. In practice, however, such a direct correlation is difficult to make since statute requirements and actual degree programs seem to conflict. For example, it is usually assumed on the basis of statute requirements that the three degrees mentioned above were granted over a number of years, since prescribed periods of study were set out for each degree. However, a majority of students at Paris received all three degrees within a single year. Henry, who in this instance is typical of the majority, determined on February 20, 1363, licensed three months later on May 20, 1363, and had been received into the ranks of teaching masters by July 18 of the same year.[6]

The problems encountered in bringing statute requirements into line with actual programs of study center on two degrees, determination and licensing. (Inception presents no problems as the required time between licensing and inception was minimal and often exceeded.) Students who were being examined at determination had to have studied at Paris or at some other *studium generale* for a period of two to three years, during which time they were required to audit a series of lectures that focused attention on the trivium. These included Porphyry's *Isagoge*, Priscian's texts on grammar, the *Barbarism* of Donatus, Boethius's *Divisions, Topics,* and *Principles*, plus portions of Aristotle's logic.[7] Students who were being examined in conjunction with the granting of the license to teach also had to have studied for two to three years, but in this case the lecture requirements were different. The texts prescribed for licensing included Aristotle's *Physics, On the Heavens and Earth, On Generation and Corruption, Meteorology, On the Soul, Brief Natural Treatises*, and *Ethics*, plus, as is suggested in one statute, "at least one hundred lectures dealing with mathematics."[8] Now, what is often assumed from these two sets of requirements is that students first pursued those subjects that led to determination and then went on to audit the lectures required to receive the license to teach. However, this simply cannot be the case. If it were, licensing would come in most cases two years after determination, and it did not; most students licensed within a few months of determination.[9] What then was the course of study that they followed?

The most plausible answer to the last question seems to be that at this point in the fourteenth century a majority of students spent only about two or three years in residence prior to determination, during which time they focused their attention on the lectures required for licensing. There are several reasons for suggesting this. First, the only resident requirement relating to either degree that seems not to have been dispensed with or that does not allow for equivalent studies at another *studium* is the three-year resident requirement for licensing.[10] Students were permitted to determine at Paris even if they audited all of the required texts at another *studium generale* and even if they had not attended for two years the required disputations and responded to questions in the schools of the nation.[11] Second, the resident requirement for licensing was interpreted by the faculty as excluding lectures on grammar (*intelligitur absque grammatica*). Moreover, most of the required texts relating to the trivium were dispensed with, either in whole or in part, at licensing. Thus, it would seem that the faculty itself realized that at licensing the greatest emphasis should be placed on subjects relating to natural and moral philosophy as set out in the works of Aristotle rather than the trivium.[12] Finally, if Henry is at all indicative of the calibre and experience of students coming to Paris, it is easy to understand why further attendance at elementary lectures may have been unnecessary. It seems hard to imagine that at thirty-eight years of age he needed to attend lectures on the elementary texts of Donatus, Priscian, Boethius, and Aristotle. Instead, he could have met all of the requirements for the master of arts degree by attending lectures on natural and moral philosophy for the required two to three years and then passing through, in very rapid succession, the customary three degrees. This would place his arrival in Paris around 1360 or 1361, not 1358 as has been suggested by some scholars.[13]

What is interesting to note from all of this is that even if it is correct to assume that an arts training at Paris in this period stressed subjects relating to natural philosophy, a student could still become a master in the arts with only three years of study, which is hardly an inordinate length of time to train a potential scientist. Moreover, once the master of arts degree was granted, only two additional years of lecturing in the arts were required before other studies in the higher faculties could be taken up.[14] Strictly speaking, then, the required years spent in the arts, and hence in science, were limited to five. Such a minimal commitment would seem to suggest that medieval schoolmen did not place a great deal of emphasis on science studies. Medieval society, through the instrument of the university curriculum, did not guide budding intellectuals into the study of nature. However, even though in general there was no strong commitment to science, there were exceptions. No maximum number of years spent in the arts was specified. As long as interest persisted, the arts could be pursued as a viable and acceptable form of study. In Henry's case this interest persisted by itself for at least ten years and, in combination with other interests, throughout the rest of his life.

Evidence supporting Henry's continued interest in the arts after 1363 comes from two sources. First, throughout the years 1363–73 he remained active in the affairs of the English-German nation and sponsored a number of students for degrees in the arts.[15] The latter activity required that he be a regent or teaching master in the arts.[16] Second, his three scientific works that can be dated with certainty (*De reprobatione, Quaestio de cometa,* and *Contra astrologos*) were written between the years 1363 and 1373.[17] Accordingly, there seems to be ample evidence supporting a fairly active career in the arts even after he completed his years of training. But the evidence from this period also suggests that not all of Henry's scientific energies were directed toward the University. He must also have had some connections with the Court of Charles V (1363–80), since it was at the latter's request that the *Quaestio de cometa* was written (*ad mandatum christianissimi regis Francorum Caroli V*).[18] If these connections were at all extensive, they may have played an important role in his intellectual development. Present at the Court during these years were such figures as James Bauchant; the Augustinian Hermit John Corbechon; John Daudin, canon of St. Chapelle; Raoul of Presles; John Golein, a Carmelite; Simon of Hesdin; and perhaps most importantly, Nicole Oresme—all of whom were engaged at one time or another in producing French translations of a wide variety of scholarly works.[19] Unfortunately, we do not know why Henry was selected by Charles to write a treatise on the comet of 1368. Not only his years of training but virtually his entire career in the arts is devoid of personal details that would give an insight into his personal contacts during these years.

The years that Henry devoted to the study of the arts more than indicate that science had a place in the late medieval world and could form an important part of an individual scholar's career. Whatever religious orientation there may have been to medieval life, it did not, at this beginning level at least, interfere with the functioning of the sciences. However, the priorities of medieval life and the assumed relative importance of various disciplines within the university community leave little doubt that the arts were not considered by most as an end in themselves. For a schoolman such as Henry, the arts were just the beginning, the jumping-off point for higher studies within the university. As such, they comprised an integral part of learning. As Henry would note later in life:

> all of the other human sciences spring from [the arts] as if from an initial fountain: medicine from natural philosophy, the two faculties of law from moral [philosophy], theology . . . from metaphysics.[20]

But in springing forth from the arts, these other disciplines leave the arts far behind in the order of the dignity of the sciences. Masters of medicine, law, and theology have the distinction of being both masters of the arts and "designated with the titles of the other sciences" (*sunt insigniti aliarum titulis scienciarum*). Compared to masters of the arts, they know "just as much and more" (*tantundem et amplius*). Therefore, to stop one's training with the arts is to stop short of climbing the ladder of the sciences to medicine, law, or the rung of greatest nobility, theology.[21] Herein lay the social pressures that eventually led Henry from his interests in the arts to a career in the Church.

The pressures of a Church career did not immediately make what was learned in the arts obsolete. Such a career frequently began with theological studies, which relied in both content and method on the background furnished by the arts. When students in theology advanced from auditors to lecturers on the two required texts for a degree in theology, the Bible and the *Sentences* of Peter Lombard, they often carried with them some of their earlier interests from the arts. Henry is no exception in this regard. His *Postilla super Isaiam* begin with a praise of the knowledge to be gained by studying creation and proceed to draw in scientific examples to help clarify the passages of text under consideration.[22] To a lesser extent, this is also true of his lectures on the *Sentences*.

Nevertheless, the fact cannot be ignored that even if science was carried over into theological studies, the approach to science in this new context was often radically different from that found in the arts. Theology students had very specific goals to pursue and, in the process, their scientific interests were often restricted. Following the completion of a degree in theology, very few masters ever returned to science and their earlier interests in the arts.

With regard to this latter generalization, Henry stands out as an obvious exception. The years he devoted to bringing an end to the Schism, to giving advice to his fellow churchmen, to reorganizing the University of Vienna, and to fostering his own interests and beliefs in theology, certainly took him far from his earlier interests in the arts. As he mentioned at one point during his Vienna lectures, the latter had occupied him earlier in life, at Paris, when he had more time.[23]

And yet, his interest in science was never totally abandoned, as from time to time it surfaces in what, at first glance, do not appear to be scientific contexts. During the Eberbach period he included some astronomical considerations in his *Tractatus de horis canonicis* and some natural philosophy in the *Tractatus de discretione spirituum.* Both of these aspects of science were later united while at Vienna in the form of an anti-astrological treatise, *Contra Telesphorum,* and in the course of his sermons on the Virgin Mary. As late as 1396, one year before his death, he engaged in a spirited defense of the arts in his *Sermo de Sancta Katharina Virgine,* a defense that could easily have been written a quarter century before when he was still a regent master in the arts and a student in theology. But in no work are his earlier interests in science more clearly expressed than in his *Lecturae super Genesim.* Here, under the guise of lectures on the Bible, Henry sets out, what is for all practical purposes, an encyclopedia dealing with the world which ultimately effects a fusion between his two life-long interests. His pursuit of science by this time has become an intensely religious activity.

2. Scientific and Science-Related Works

As the preceding biographical survey should make clear, Henry's life and writings divide into two distinct phases. Prior to 1374 there is no direct evidence of any interests other than those relating to the arts. All of his writings that antedate this year are scientific in content and show no influence of other studies. Following 1374, all of his extant and datable writings have nonscientific foci. Although many of these later works contain discussions of scientific topics, none has what can be considered an arts orientation. This leads to a natural division of the works that have bearing on his science into two categories: "scientific works" and "science-related works."

A. Scientific works

Of the many treatises that Henry wrote of a purely scientific nature, three can be dated precisely: *De reprobatione eccentricorum et epicyclorum* (1364), *Quaestio de cometa* (1368), and *Tractatus contra astrologos conjunctionistas de eventibus futurorum* (1373).[24] The first is a short astronomical treatise in which Henry sets forth a number of objections to Ptolemaic astronomy along with a brief discussion of an astronomical system that offers an alternative means of accounting for celestial phenomena.[25] The latter two are basically anti-astrological in tone; the former argues against predictions from comets while the latter rejects predictions from conjunctions. *Contra astrologos* closes with a brief chapter on the origin of events in this world, which assumes that they are not directly caused by the stars.[26]

An internal reference allows one more work to be dated to the period prior to 1374. In *Contra astrologos* Henry says that he has discussed a particular point "*in tractatu quodam de natura communi*," which is undoubtedly a reference to *De habitudine causarum et influxu naturae communis*.[27] Since this work is related very closely in content to *De reductione effectuum*, the latter is usually dated to the pre-1374 period, although this cannot be proven directly.[28] *De reductione* takes up the study of deriving causes from their effects, whereas *De habitudine* focuses attention on celestial causes and the manner in which they produce specific effects.[29]

The remainder of Henry's scientific works are undated and undatable by means of internal references. This leaves no choice but to rely on their content to place them within the context of his academic career. Fortunately, two factors stand out that allow for a fairly accurate description of the order in which his works were written: their style and their emphasis on astrology.

Stylistically, most of Henry's scientific works are nonscholastic and polemical. In only two major works, *Egregia puncta et notata de anima* and *Quaestiones super perspectivam,* does he utilize the *quaestiones* form of argument usually associated with scholastic science.[30] His remaining works dispense with the formal structure of the *quaestiones* style and follow instead the book-chapter pattern of organization more commonly associated with nonscholastic science.

In addition, when clarifying arguments and dispensing with contraproductive lines of reasoning in his polemics, his interests are obvious; he is strongly oriented against the evils of astrology. Anti-astrological sentiments are not, however, found in all of his works. In fact, the contrast between *Quaestio de cometa* or *Contra astrologos*, in which the problems of astrology and superstition are referred to again and again, and *De reprobatione*, leads to the suspicion that his obsession with astrology may have developed in the course of his lifetime and was not a constant feature of his thought.

Putting these two factors together, a scheme emerges for dating his scientific works. If it is assumed, following statute requirements, that the most likely years for lectures in the *quaestiones* style would be the biannum immediately following the granting of the master of arts degree (1363–1365), then *Egregia puncta et notata de anima* and *Quaestiones super perspectivam* can tentatively be dated to these years. Neither work has even the slightest hint of any preoccupation with astrology and so would tend to confirm, following the 1364 date for *De reprobatione*, that early in his career Henry's interests in astronomy were more purely technical and mathematical. All three works provide numerous opportunities to launch into discussions of celestial influences, none of which is ever utilized for that purpose. Henry's interest in his early years in technical astronomy is further supported by his noting that even before writing *De reprobatione*, he had written another work "*de motibus planetarum secundum eccentricos et epicycles*."[31] Finally, if it is true that Henry's early interests did run toward technical astronomy, one additional work, which provides a very elementary ennumeration of astronomical terms along with brief definitions, can be added to the list of early works–a work entitled *Expositio terminorum astronomiae*.[32]

The common core of interest expressed in this grouping of presumed early works is significant. The *Quaestiones super perspectivam* rely heavily on astronomical examples and even make the point that astronomers must be experts in perspective if they are to properly understand their observations.[33] The *Quaestiones* also show a pronounced interest in psychological matters, thus relating them to the *Notata de anima*.[34] Both the *Notata* and *Quaestiones* have long discussions of light. If these parallel interests are combined with the fact, already mentioned, that in none of the presumed early works is astrology given even a slight rebuff, despite numerous opportunities, it does seem likely that all were composed before 1368 and probably in conjunction with his duties as a lecturer in the arts.

Just as the Schism may have diverted Henry's attention from scientific interests to matters relating to the Church, so too, the comet controversy of 1368 and Henry's subsequent entry into disputes over prognostication may have turned his attention away from science *qua* science to science as it applies to astrological concerns. At this point his style shifts from that of a school lecturer to that of a polemicist whose aims now extend beyond the classroom.

The *Quaestio de cometa* quite nicely exhibits this shift. It begins by raising the one question that serves as the basis for the entire treatise–that is, "whether the appearance of a comet would be a sign foretelling certain events?"[35] After dispensing with this question in scholastic form, the remainder of the work shifts to the topic-by-topic, chapter-by-chapter style that

Henry's concern with science is evident throughout the *Lecturae*. He seems always to have had scientific information close at hand in case an occasion should arise where such information might apply. No connection with nature, however obscure, seems to have been passed over without at least a brief mention of the principles or controversies involved. Coincidentally, his closing lecture ends in much the same way as he had begun some six or more years before—that is, discussing the heavens, orbs, and celestial movers.[51]

Nevertheless, there can be little doubt that the lectures that have the most bearing on science are those dealing with the six days of creation in chapter one of Genesis. It is here that Henry's dictum—"he who does not know or turns away from the cleverness of the working craftsman or the subtlety of the artistic work is amazed less by the wisdom of the craftsman"[52] —has its most direct application. Accordingly, it is on these lectures that I have focused my attention in this book, bringing in related material from the remaining lectures and from Henry's other scientific works as the occasion warrants.

Including science in a discussion of the six days of creation is not new to the fourteenth century. Hexameral literature has a long history that extends back at least to the Fathers of the Church, both Greek and Latin.[53] In one sense it might be argued that science and creation have always gone together, as is amply evidenced by the Platonic tradition stemming from the *Timaeus* and the primitive creation stories contained in mythopoeic literature.[54] The Latin Middle Ages were not strangers to this tradition. During the early Middle Ages the Venerable Bede, Isidore of Seville, Walafrid Strabo, and Rabanus Maurus all included long scientific discussions in their treatment of the six days, as did the Platonists of the twelfth and early thirteenth centuries: Bernard Silvester, Thierry of Chartres, William of Conches, Honorius of Autun, and Robert Grosseteste.[55]

However, with the rise of Aristotelianism in the thirteenth and fourteenth centuries, scientific discussions shift from the context of creation to the context of the Aristotelian corpus; and as this happens, hexameral literature takes on more theological tones. Aquinas's discussion of the six days in his *Summa*, for example, contains very little technical science and stresses more metaphysical and theological concerns.[56] The same would be true of discussions of creation set out in commentaries on book two of the *Sentences*.[57] Thus, in including so much science in his discussion of the six days, Henry is in some ways harking back to the Platonic tradition of the twelfth century, which he draws upon very little, and to the earlier precedent set by Augustine in *De genesi ad litteram*. In doing so, he is certainly following the general outlines of the Augustinian revival of his day,[58] although he seems to have been the only fourteenth-century Augustinian who followed the revival to one of its most natural consequences, the pursuit of science within the context of creation.

Chapter one of Genesis provides more than a simple excuse for discussing science; it also provides a convenient and familiar outline for pursuing a detailed analysis of the world, proceeding as it does from higher and more ethereal to lower and more earthly things. Its familiarity stems from the fact that not only other Genesis commentaries but most of the encyclopedic tradition of the Middle Ages follows a similar outline. A comparison of the contents of Pliny's *Naturalis historiae*, Isidore's *Etymologiae*, Vincent of Beauvais's *Speculum*, or Bartholomew the Englishman's *De proprietatibus rerum*, with the *Lecturae* would show a remarkable similarity in the order in which topics are treated.[59]

Its convenience stems from the fact that within the context of the six days, virtually no aspect of science is ignored. The work that precedes the six days *in principio* touches on metaphysics and its related theological and physical concerns. The creation of heavens and earth brings in astronomy, astrology and the other occult arts, cosmology, cosmography, meteorology, and chemistry (the generation and corruption of the elements). The closely related activity of producing light has bearing on optics (perspective). Then, as God ornaments the elements with plants, animals, and man, the related biological sciences come under consideration along with psychology and medicine.[60] As Vincent of Beauvais noted earlier in setting out a similar plan of discussion, "from the beginning of the creation of things to the rest of the Sabbath . . . are found those things that pertain to the nature of the heavens and the earth."[61]

Had Henry followed this straightforward, day-by-day division of the first chapter of Genesis in his lectures, his ideas on creation and the world could be reconstructed with little difficulty. He did not, however. To begin with, he initially divides the six days into two major subgroups: days one through four, and days five and six.[62] The first of these subgroups is further divided according to four different approaches, following the opinions of two famous literal expositions (*expositiones famosiores et solemniores atque propinquiores sensui litterali*) and three established opinions on the nature of the heavens (*tres . . . opiniones solemnes*).[63]

These expositions and opinions are dealt with in the following manner: (1) There is a discussion of the opinion of the first famous exposition, which in turn examines the first two established opinions (*Lect.* 1:13ra ff.). This section of the *Lecturae* follows Nicholas of Lyra's *Postilla* very closely.[64] (2) Then, the second exposition, which is followed by Bede and Strabo, and the third established opinion are discussed (*Lect.* 1:20ra ff.). This section begins with verbatim quotations from the *Postilla* but adds a great deal more material.[65] (3) Henry summarizes the opinions of the first two expositions and the import of the three established opinions with an eye toward clarifying how God instituted creation (*Lect.* 1:46va ff.). (4) Finally, he goes on in a fourth way (*quartomodo*) discussing days one through four (*Lect.* 1:53ra ff.), which leads to part five (*Lect.* 1:78va). At this point the multiple approach to the text is abandoned and the discussion of day five begins.[66]

The significance of these further divisions of the *Lecturae* should not be overemphasized. At no time does Henry ever specifically limit himself to a single point of view; most often he uses the opinions of the prior commentators, who for the most part remain nameless, to introduce topics that are to be pursued in greater depth. Therefore, it is usually of little consequence that at some point a topic is being expounded according to one particular exposition, since most of the ideas being expressed are Henry's. What is of consequence, however, is the fact that this further division of the lectures spreads the discussions of astronomy, physics, and so forth, throughout a number of lectures. It is essential to realize this point since the unity that is presented in the chapters that follow is mine and not Henry's. Without an index, there is no way to predict at what point in the *Lecturae* a particular topic will be raised or whether a single discussion of that topic will be repeated or expanded in another lecture.[67]

3. The Historical Perspective of the Medieval Scientist

The opinions drawn from the three famous expositions represent only a small fraction of the authorities quoted or referred to in the *Lecturae*. Unlike his earlier scientific works, which remain fairly narrow in the scope of their coverage and in the number of authorities cited, Henry ranges broadly in the lectures on the six days, mentioning well over fifty authorities and numerous intellectual traditions. These citations are important because they provide an initial insight into the intellectual heritage within which he falls and—the topic I am initially concerned with—his attitude toward the past. What I am after in this latter instance are not Henry's specific statements about authority and its historical context, but rather his assumed and unarticulated vision of the past as it functions in the normal course of what are essentially theological-scientific lectures. Medieval schoolmen had more authorities to choose from than they cared to or probably ever could have used. Their historical vision extends from their immediate intellectual milieu back at least to the pre-Socratics or, occasionally via conjecture and inference, to the beginnings of speculative thought among the Egyptians, and covers all of the terrain that lies between. How does this tradition sort itself out in the course of Henry's lectures?

Henry's vision of the past can be divided into three distinct elements: (1) Most obviously he sees stretching before him a succession of authorities whose thought in aggregate comprises a chain that connects past to present. Time, in this instance, flows from individual to individual, from Plato or Aristotle through late Latin figures and Arabic writers to the Latin tradition of his own day. Breaks in the chain occur where there is disagreement between individuals. Augustine may disagree with Aristotle, Boethius with Plato, and so forth. And eventually, from this flow of time come certain preferences that explain where Henry's own thought resides within the intellectual heritage of the past. (2) On a more general level, as thinkers merge into schools, Henry sees breaks in the chain that connects past to present. Platonists and the Holy Doctors are ancients as opposed to the moderns of his own day. Sometime between the distant past and the immediate present a division occurs that sets out contrasting ages. Even without the Renaissance concept of a "middle age," medieval thinkers were aware of the gap that separated their own age from the past. (3) The feeling of separation gives way on yet another level to a deeper feeling of continuity. Even though at times Henry feels alienated from the tradition of the ancient past, more frequently he speaks of the past as though it stretches without major breaks to the most recent of moderns.[68] Ultimately, his own work is tied via disciplinary traditions to a common heritage that extends to the primitive beginnings of each of the sciences.[69]

What is meant in this instance by disciplinary traditions are the frequent references to groups of thinkers who are, for one reason or another, lumped together under a single term: philosophers, Peripatetics, and so on. Throughout the *Lecturae* Henry mentions such groups at crucial points in his arguments, usually for the purpose of providing a note of authority or clarity (for example, such and such is true according to the "mathematicians," or something is referred to as a theory held by the "physicians").[70] The strength of such citations is that they relate the issue under consideration to a tradition that extends over long periods of time. Philosophers and Peripatetics represent a common mode of thinking that extends from Aristotle to the fourteenth century, theologians and Catholic Doctors one that extends from the early centuries after Christ to the present. The same would be true of astronomers, gentiles, wisemen, poets, rational doctors, astrologers, mathematicians, physicists, meteorologists, alchemists, and natural philosophers.[71] But more than simply supplying an element of extended time, the disciplinary tradition also directly ties the present to the past. As a natural philosopher, astronomer, or physicist, Henry has direct ties with the ancients. What they did and what he is doing are in many ways the same. On this general and important level, past and present very much fuse in his thought.

It is only on more particular levels that the fusion between past and present breaks down. On specific issues Henry does recognize that the positions held and methods used by his contemporaries may not correspond to those of the past. In reference to a discussion of first matter, for example, he notes that Hugh of St. Victor's argument follows "the opinion of the ancient philosophers" (*secundum illam veterum philosophorum opinionem*), an opinion that is not held by "modern philosophers" (*secundum modernos philosophos*).[72] So there are differences of opinion; this much is obvious.

What is not obvious is where the break between ancients and moderns occurs. "Ancients" certainly include the Greeks (Plato and Aristotle) and the Fathers of the Church.[73] They can also include Arabic authorities.[74] "Moderns," on the other hand, is sometimes used broadly to refer to the Latin tradition of the West[75] or, more narrowly, to contemporaries, although Henry seldom uses modern in the latter sense. When he wants to refer specifically to his own times, he usually uses precise terms such as "the Parisian Peripatetics," "the present astronomical tradition," or "our doctors."[76] In general, then, Henry's use of the terms "ancients" and "moderns" is not precise. Where Bede and Strabo stand, for example, is never made clear. What is clear, however, is that the split between ancients and moderns does not undercut Henry's feeling that the intellectual tradition within which he is working is one of long-standing duration. In the final analysis, it is his ties with the ancients, not his departures from them, that are most important.

The importance of Henry's ties with the past becomes even more obvious as we come to the third aspect of his historical perspective, individual authority. Here, his use of contemporaries pales in comparison to his stress of the ancients. From among fourteenth-century writers, only Guido Boneti, Nicole Oresme, and Nicholas of Lyra are cited by name in the lectures on the six days of creation.[77] As we go further back in time, the number of authorities cited and the frequency of these citations increases proportionally, as is obvious in the following listing of Henry's sources (I have starred [*] important sources of twenty or more citations, and double starred [**] critical sources of one hundred or more citations):

Latin, 12th and 13th C. Accursius, Alain de Lille, *Albert the Great, Alfonsine Tables, Anselm, Thomas Aquinas, Bernard of Clairvaux, the Decretalists, Giles of Rome, *Hugh of St. Victor, Joachim of Fiori, John Pecham, Richard of St. Victor, Sacrobosco, and William of Auvergne.

Latin, Early medieval. Venerable Bede, *Isidore of Seville, Rabanus Maurus, and Walafrid Strabo.

Arabic. al-Battani, Albumasar, Alfarabi, Alkindi, Alpetragius, Averroes, Avicenna, Geber, Haly Abenragel, and Mohamet.

Jewish (probably cited through Nicholas of Lyra). Rabi Eliser, Rabi Salomen, Thābit ibn Qurra, and the Talmud.

Roman. Boethius, Martianus Capella, Cassiodorus, Diogenes, Pliny, Seneca, Ticonius, Valerius Maximus, and Virgil.

Early Christian. **Ambrose, **Augustine, Basil, Chrisostomus, Jerome, John Damascene, Gregory of Tours, and Origen.

Greek. Anaxagoras, Archimedes (pseudo), **Aristotle, Calippus, Democritus, Empedocles, Euclid, Eudoxus, Hippocrates, Plato, and *Ptolemy.
***The Bible.*

It is upon Augustine, the Bible, Aristotle, and Ambrose, in that order, that Henry ultimately places the most weight, even in the portions of his *Lecturae* that are almost entirely on science.

Taking note of the authorities upon whom Henry places the most weight gives a preliminary insight into the orientation of his thought. Citations alone do not, of course, give a complete picture (more will be added later as particular ideas are discussed). But apart from ideas, we can get an impression of his historical tastes. On the basis of citations, Henry's historical leanings seem to be those of someone who has been thoroughly trained in the Aristotelian world view and who has come to interpret that world view through the eyes of the neo-Platonic tradition that extends from Augustine through Hugh of St. Victor, Albertus Magnus, and Giles of Rome to the important Augustinian thinkers of the fourteenth century.

Neither of these leanings (Aristotelian and Augustinian) should be at all unexpected. Despite the occasional critiques and modifications of Aristotle's thought that took place over the course of the thirteenth and fourteenth centuries, the orientation of late medieval science remains solidly Aristotelian. Most of the important developments in science that took place over the course of the fourteenth century were made within the Aristotelian world view and served to strengthen rather than weaken it.[78] Moreover, by 1385, when he began lecturing on the Bible, Henry had already given indications of his close ties with the Augustinian (Cistercian) school at Paris. His own *Sentence* commentary, composed at Paris, relies on the *Sentences* of the Augustinian, Ficinus of Ast.[79] In addition, his close friendship with James of Eltville and his rereading of James's *Sentences* at Eberbach certainly suggest that he was more than open to the ideas of this school. His preference for Augustinian-Peripateticism seems, therefore, to be very much a part of his life as outlined at the beginning of this chapter.

In noting these preferences, finally, one point needs to be borne in mind. Preference should not be interpreted as unthinking devotion. What it means is that schoolmen, such as Henry, in selecting the environs in which they would ultimately work, sought out the authorities within whose philosophical and theological purviews they would feel the most comfortable. Just as later thinkers would become Cartesians, Hegelians, or Comtians, and modern scientists become logical positivists or experimentalists, so too, Henry sought out the workshops of Augustine and Aristotle in which he would carry out his task. In the process, of course, he assumes the intellectual biases of these thinkers. The authorities he reads, the sources he quotes, the methods he uses—all bear to some degree the mark of his day and the influence of his sources. But this does not make his mind less original than that of the twentieth-century scientist who is tied to mathematical suppositions and experimental methods. It simply makes his approach to science radically different from our own. This different approach and the world view that results from it are the central themes to which we now turn.

CHAPTER II

In principio: Matter, Form, and Metaphysics

*In the beginning God created the heaven and the earth. The earth was empty and void, and darkness was upon the face of the abyss. And the spirit of the Lord was borne upon the waters. Genesis 1:1–2.**

Creating an artistic masterpiece such as a clay or marble statue involves drawing forth an imagined form (a human figure, a centaur, whatever) from some medium or matter (clay or marble in this case). Creating the universe involves drawing forth a form from nothing, *ex nihilo.* As the artisan who created the universe, God faced in the beginning a problem that no other artisan has ever faced; he had to create both the forms that would allow his masterpiece to come into being and the matter from which those forms could be educed, and he had to do this *ex nihilo.*

Whether or not creation *ex nihilo* is within God's capability is not at issue in a discussion of science and creation. Henry concludes at the beginning of his *Lecturae* that God can create *ex nihilo,* and to doubt such would divert our attention from science to the domain of theology.[1] What is at issue is precisely how God set about this difficult task. How did he bring forth plants and animals from the lifeless elements? How were the heavens divided into stars, orbs, and intelligences? What forces and connections did he establish between the heavenly and earthly realms so that the chain of events in this world would run smoothly? What dispositions were originally given to the universe so that provision was made for all that was to follow? And how, in the very beginning, did he create something *ex nihilo* that would serve as the foundation for being through the entirety of its existence?

This last question is the first that needs to be considered, the question of "*in principio.*" At the very outset of creation God brought into being both the forms and matter needed to create the universe. Both had to be equal to the task they were to perform; there had to be inherent in them the entire design of the masterpiece that was to follow and the capacity to become that masterpiece as the events of creation unfolded. *In principio* provision had to be made for the heavenly and earthly realms, for corporeal and incorporeal being, for the hierarchical perfection that extends from men to angels, for the capacity of things to change in part while remaining in part the same–in short, for all of the myriad details that are manifest in the world in which we live. Both forms and matter also had to be equal to the task while being "empty and void," without any identifying characteristics. *In principio* there must be everything and nothing. How could this be and how did God begin the task of creating the universe?

1. The First Instant

Knowledge about creation rests initially on knowledge about the Creator. Without some information about how God proceeded *in principio*, all subsequent speculation about creation is futile. This is because the God of creation is an omnipotent God, one who "is able to act absolutely, in whatever way and however, and to have for himself whatever" he desires (*ipse solus quodcumque, in quocumque, et qualitercumque absolute agere potest vel se habere ei*).[2] He could, despite Aristotle's arguments to the contrary, create a plurality of worlds, thereby making it impossible for us to know whether our world with its laws is an isolated entity or the totality of being.[3] His actions in this world could be arbitrary and direct, thus by-passing the normal course of nature. God could, as one commentator suggests, make day and night by simply turning on and off the lights of the heavens, allowing them to radiate for the duration of the day and stopping them for the duration of the night.[4] He could, in short, do anything at anytime. Happily, however, he does not.

Despite the tendency of fourteenth-century thinkers to stress God's absolute power, few allowed his omnipotence to undermine their belief in a natural order.[5] This is due to the fact that creation is not so much the product of omnipotence as it is the product of a chosen course of action. *In principio* God selected from the infinite choices and possibilities open to him the finite number of things that inhabit the universe (*de infinitis creaturis possibilibus fecit solum paucas*).[6] In his wisdom he saw that it was best to create one finite world, not one infinite world or many finite worlds. He also saw that it was best to populate this world with a given number of entities, including both good and evil things, and to establish in the world a given and natural mode of action.[7] This is not to say that God cannot or, upon occasion, does not transcend the natural order. As will be noted later (chap. VI, sec. 1), Henry allows for two levels of causation, natural and supernatural, and believes that occasionaly the laws of nature are suspended. But exceptions in this case only serve to prove the rule; they do not indicate that in every case, or even in many cases, God suspends the laws of nature and acts through supernatural causes. In fact, supernatural events are so rare in the natural order that Augustine suggests with regard to Scriptures that "we ought to seek to understand how God instituted the natures of things, not how he may have wished to work by way of miracles. . . ."[8]

Understanding creation begins, therefore, by accepting the supposition that in a given instant of God's eternity, the instant entitled *in principio,* he freely chose to bring into being *ex nihilo* one finite world that follows a single and generally consistent mode of action.[9] This is his initial creative act. As such, it is unique; first, because it is the only act that involves creating *ex nihilo* and second, because it takes place before time and apart from time.

Time is the measure of motion, the measure of something progressing from one point to the next point. Prior to the beginning there was no point from which to progress. Prior to *in principio* nothing had been created, so creation *ex nihilo* is done apart from time, while there is duration but not time (*ibi fuisse prioritatem durationis et non temporis quia tempus praesupponit motum successivum*).[10] Time begins with motion, when God says "let light be made."[11] Or to put the issue in a slightly more graphic form, *in principio* stands as the dimensionless point that begins temporal being. As a point, it has no dimensions, nothing that permits it to be identified. Succession, which time involves, cannot be made up of a single point. Succession occurs when we add a second point, and this happens only after "God created the heaven and the earth."

The dimensionless first instant that stands at the beginning of creation occasioned a great deal of debate among Biblical commentators. What was at issue was the amount of the work of creation completed in this first instant. Augustine, who in this case followed a line of reasoning set forth by earlier Greek commentators, developed the theory of instantaneous creation—that is, that all of the work of creation is done in one initial instant and not over the course of six temporal days. As Henry summarizes this position, it admits a priority of nature (*prioritas naturae*) but not a priority of time (*prioritas temporis*); in the order of nature, solar light precedes the light of the moon, but both are created in the first instant. The obvious consequence of this view is that the six days cannot be interpreted literally. Instead, they must be assigned an allegorical interpretation, as is the case with the Augustinian view of creation which posits that the terms "morning" and "evening" are best understood as different ways of coming to know the creative act.[12]

A more common interpretation of the work of creation in Henry's day takes the account given in Genesis literally, as signifying six days of twenty-four-hour duration. The first instant is then separated out from the six days and assigned what is narrowly referred to as "the work of creation" (*opus creationis*). Thereafter, according to this second interpretation, this work is followed by two further creative acts, the work of distinction (*opus distinctionis*) and the work of ornamentation (*opus ornatus*). During the former, heaven and earth are distinguished, first into day and night (day one) and then into the individual heavenly spheres (day two) and earthly elements (day three). During the latter, the universe is ornamented, beginning with the heavens (day four), and then extending to the middle elements: water, air,

combining with one another, these bodies produce the changes that we observe in the world.[20] (2) Anaxagoras (fl.460 b.c.) grounded the seemingly infinite number of things that surround us in an infinite mixture (*per infinitam mixturam*). That is to say, rather than positing that things are made up of simple bodies, he reduced each thing to a principle, thus reaching the conclusion that there are as many principles as things, or seemingly an infinite number of principles.[21] (3) Plato reduced atomism to mathematics. In the *Timaeus* he describes things in geometrical terms, as being composed of surfaces and dimensions. When things are reduced to their simplest form, number evolves as the basic constituent of the universe; bodies are made up of surfaces, surfaces of lines, lines of points or numbers.[22] There can be no doubt, therefore, that atomism as a system for explaining the foundation of the universe was known in medieval times. And yet, with very few exceptions, it was neglected—even more than it was rejected—as a viable alternative to the prevailing philosophies. Henry's response in this case is characteristic of his age. He passes over these theories without even a summary dismissal.[23]

It is sometimes suggested that the neglect of atomism by medieval scientists has its origin in the rigidity of medieval thought.[24] Certainly there is some truth to this suggestion; however, there is a far simpler explanation of its sparse showing in history prior to the seventeenth century. For a premodern scientist such as Henry, explaining the complex operations of the world with invisible and undetectable bodies is simply beyond his ken. Atomism might work well when it comes to explaining simple phenomena such as solution, burning, rarifaction, and condensation (as indeed it had for Plato), or even some rather simple psychological activities as set out by later atomists.[25] But what about more complex phenomena: growth, life, thinking, and so forth? How are they to be explained? How, for example, using only particles and motion, can one hope to account for the many changes that take place when an acorn grows to an oak tree? What laws of motion can account for and order the countless changes needed to explain growth? What assurances can be given that as atoms come together, a tree and not some other object will ultimately emerge? In short, what mechanisms does atomism provide for bringing design—the complex design of everyday experience—into nature?

On a working level it provides very few. Prior to molecular biology, physical chemistry, nuclear physics, and so forth, the number of ways that scientists had to account for the design of nature through atoms were limited. There were just so many ways to imagine small bodies and motion before the imagination was stretched beyond the bounds of credibility. Therefore, what is so puzzling, from a historical point of view, is not that so few persons discussed atomism prior to the advent of modern science, but that those who did retained their belief in atomism even when faced with its fanciful consequences.[26]

But if atomism as a philosophy for explaining complex phenomena does not work, what does? The answer to this question rests on solving the problem of design. What needs to be found is a way to explain how, as change takes place, one object emerges, our oak tree for example, and not some other object. One way would be simply to assume that design is somehow imposed from the outside, that nature, as its actions unfold, is constantly under the guiding hand of an influence that keeps the entire system in proper order. This is the assumption that lies at the heart of mythopoeic thought and emerges at times in the writings of later figures, such as Isaac Newton.[27]

There is, however, another way to resolve this problem, one that does not regularly relegate design to the influences of an outside agent. This would be by putting design into nature, by making design a natural and normal part of the functioning of the world, which is precisely what medieval scientists, as Aristotelians, did.[28] Rather than assuming that the things we see are the manifestation of innumerable invisible bodies occupying various positions, they posited that what we see are shapes and configurations—called "forms"—that express themselves at the time a particular object comes into being. The acorn becomes a tree not because trillions of atoms somehow miraculously come together to form a tree, but because, over time, the form of the acorn is replaced by the form of an oak tree. Design as form is a part of nature; growing, thinking, and the other complex phenomena of everyday experience proceed as they do because their course is established in nature.

It is assumed, therefore—as we return to our initial problem of what God did *in principio*—that in the beginning he began the process of bringing design as form to the universe (*in aliquibus proprietatibus essent informia*).[29] Creation can be defined as the process by which forms, which make things what they are, are transferred from the Creator to what will become the realm of things. This definition raises one further question. Forms are outlines of what something will become; they are not themselves anything. They possess the capacity to shape something, but until they shape something, they are nothing. (Henry does not adhere to the Platonic doctrine of divine ideas.)[30] But what do forms shape; in what do they reside? The answer to this question is *in something,* a something called *first matter* or, more simply, *matter*—hence, the *matter-form* theory of the universe. Since this theory is central to virtually everything discussed in the remainder of this book—that is, to the whole of medieval science—"let me say a few words about this for those who have not attended sufficient lectures on natural philosophy" (*non audiverunt sufficienter naturalem philosophiam*).[31]

Of the two parts of matter-form composition, matter is perhaps the more elusive. As Henry notes in his *Sermo,* it is easier (*evidentius*) to prove that there is one God than that there is first matter.[32] Nevertheless, some cautious speculations about its nature were commonly thought to be within the realm of possibility.

Since nothing exists prior to creation, matter must come into being *ex nihilo*. Henry very specifically and sharply dismisses the alternative to this assumption–the view that there exits apart from God some eternal matter or *hyle* that cooperates in creation. This position, he argues (in terms that hardly allow for debate), is rejected by both philosophers and theologians.[33] In addition, since matter is receptive of forms, it must be capable of receiving forms. Henry would agree that matter is passive.[34] Yet the question remains; how passive is it? Is it stripped of all responsibility in directing the matter-form union? Commonly his contemporaries (*secundum modernos philosophos*) assumed that it is, arguing that matter is a "simple, purely passive subject" (*quaedam simplex subiecta pure passivae potentialis*) that has room (*capax*) for all forms.[35] However, Henry is not willing to go this far. Rather than assuming that hylomorphic composition, the product of form informing matter, results from the union of two extremes, he views the process of form uniting with matter as a cooperative process. Not only must two things coexist and be of diverse species for them to come together, one must have an aptitude for the other and the reverse (*"a" habeat aptitudinalem ordinem . . . in "b" et econverso "b" in "a"*).[36] Matter must have in its potency an aptitude for, or inclination toward, the forms that it will ultimately receive. There must be in it seeds or seminal bodies (*corpora quasi seminalia*) that cooperate in the matter-form union.[37]

Henry's belief in the existence of seminal dispositions in the potency of first matter is undoubtedly influenced by the Augustinian doctrine of seminal reasons.[38] However, in maintaining his own version of this doctrine, he is careful not to incline too far toward what he refers to as the ancient doctrine of "diminished being" (*esse diminutus*). The latter held that first matter has forms in it that are reduced in such a way as to be imperceptible (*latitare*).[39] These forms do not overcome the chaos of *in principio* because they have yet to express themselves. They express themselves only over the course of the six days of creation, gradually assuming in the process the recognizable shapes of everyday experience.

Henry's conception of *seminales* does not go this far toward maintaining the actual existence of forms in first matter. He posits, instead, that there is in first matter a "natural aptitude and potency" (*potentia et aptitudo naturalis*) that inclines matter toward certain forms and away from others.[40] Or to put the issue in a slightly different way, the seeds in first matter are intrinsic potentialities toward form but are not forms of themselves (*intrin-*

seca potentialitas ad formas diversas relata).[41] As a consequence, when Henry describes the unformed chaos of *in principio*, he pictures it as being devoid of forms, but as having the seeds (*seminales*) of or inclinations toward forms distributed throughout in roughly the same order that unformed matter will assume once it is distinguished and ornamented.[42] The design of creation is potentially present, even in the first instant.

What will eventually change the potential design in matter into the design of nature is, of course, form. Form is the active principle that actualizes potency. It makes a thing what it is. But what it is is fairly complex. There are various ways in which a thing can be said to be a thing, and so form, unlike matter, needs to be further resolved. A moment's thought about our oak tree will make clear why this is so. If we observe an oak tree over a number of years, it both changes and remains the same. Regardless of the season or the passing of years, as long as the tree remains alive, we can always observe that there is an oak tree standing in the front yard. And yet, at one time it was small, now it is large; at times it has green leaves, sometimes the leaves turn brown and fall to the ground. In short, the things that we observe in everyday life seem to change at the same time that they remain the same. Something vital to the tree, its *substance*, or what makes it an oak tree, remains constant over many years. Other aspects of the tree, its surface features, or what were commonly referred to as *accidents,* constantly change. To account for this duality, two types of forms are needed: one is the substantial form (*forma substantialis*), which turns matter into a particular thing, an oak tree, an element, whatever; the other is the accidental form (*forma accidentalis*), which brings to the substance of the thing its observable characteristics–whiteness, hotness, coldness, and so forth.[43]

To summarize, the essence of the matter-form theory as it will eventually be adopted by Henry and as it was commonly used in his day, is its adherence to a threefold description of things: matter, substantial form, and accidental form. Matter underlies the universe and serves as a passive, although potentially ordered, site for the activity of forms; substantial form makes objects into particular things; accidental form characterizes a thing as that thing.

It is important to note at this point that of the three components of being, one (first matter) cannot be perceived and may not even be knowable, while the second (substantial form) is only indirectly knowable through the third (accidental form).[44] There is, in short, not an abundance of direct evidence to support the complexity of the matter-form theory. The lack of direct and universally accepted evidence for this theory led to a great deal of debate among scholastic scientists. Under the guiding principle that hypotheses should not be multiplied without necessity,[45] alternative theories were put forth that dispensed with various ingredients of the matter-form world in favor of simpler components. Henry describes three such theories.

The first, a four element theory, denies the Peripatetic distinction between substantial and accidental forms (*negantes transmutationes accidentales fieri secundum formas distinctas a substantiis*).[46] According to this theory, things are made up of four simple principles (*quatuor prima simplicissima rerum generabilium et corruptibilium principia*) that inhere directly in a basic subject or substance. The principles are themselves changeable (*sint variis modis transmutabilia*—presumedly fire can become water) but their subject is not (*secundum substantiam naturaliter sunt incorruptibilia*), thus leading to the assumption (mine) that all things are substantially the same.[47] In other words, the principles of the elements (earth, water, air, and fire), which are transmutable, inhere in a subject or substance (a basic stuff), which is not transmutable. Consequently, the changes that we observe in the world are the result of the four elements combining with one another to produce various combinations.[48]

The second theory concurs with the first in having one basic substance or subject in which all forms inhere (*substantialiter eiusdem rationis*), but denies that these forms are limited to four in number. A thing becomes a thing when the common subject (*commune subiectum*) takes on a particular form or complex of forms. Change takes place when one group of forms leaves (*desiniunt*) and another enters (*adveniunt*) the common subject.[49] When the oak tree grows, the form of the acorn departs—it is not destroyed (*nihil naturaliter corrumpitur*)—as the form of the oak tree enters the common subject. Again, the distinction between substantial and accidental form is denied.

The third theory rejects matter-form composition altogether, positing, instead, that when a thing comes into being, everything about it comes into being, even its substance. As a result, there are, according to this theory, as many substances as things, or, to reverse the argument, all things are substantially different (*res corporales sint substantialiter diversarum specierum*). The consequences of destroying the substantial underlying permanence in change, which the first two theories do not do, is potentially quite radical. If nothing remains constant throughout change, then when "a" becomes "b", "a" is totally destroyed and "b" is created *ex nihilo*. All connections are broken; contingency is potentially destroyed. However, this theory also rejects the notion that total annihilation and creation *ex nihilo* accompany change (*talis corruptio non esset annihilatio . . . nec talis generatio esset creatio*). Since "a" does follow "b", there is some connection—a connection that is left unclarified except for mention of the example of the Eucharist.[50]

Although Henry gives no information about the origin of the three theories, the general focus of each allows some suggestions to be made. The four element theory obviously has a long history that extends to the writings of the pre-Socratics, as reported by Aristotle.[51] A less than critical reading of the works of either Plato or Aristotle would add these two names to its list of supporters, as was commonly the case in alchemical writings where it is used to explain transmutation.[52] Henry's immediate source for the four element theory is somewhat more puzzling. He notes at one point that it is "followed by Bernard" (*sequitur Bernardus, ille qui illudit Teutonicis*).[53] It is my feeling that this reference is to an alchemical treatise falsely circulating in Henry's day under the name of Bernard of Clairvaux (1090–1153). My reason for suggesting this is that the only other time a Bernard is mentioned by name in the discussion of creation, it is with reference to an alchemical matter and under the expanded title "*Beatus Bernardus.*"[54] Just which of many spurious and untitled alchemical tracts Henry may have had in mind, I do not know.

As to the second and third theories, their somewhat radical nature tends to suggest that Henry is referring to a debate that had taken place over fifty years before between the well-known critic of Aristotle, Nicholas of Autrecourt (fl. 1320–50), and a certain Giles.[55] At issue between the two in the section of the debate that has bearing on the matter-form theory is whether it is possible to prove that anything underlies change. Giles suggests that it is, assuming in the process, that, to quote Nicholas,

> natural change is the acquisition of a certain thing in a certain subject, with the destruction of the prior thing in the same subject.[56]

This description could easily be construed, following the second theory, as visualizing change as the coming and going of form in a common subject.

In rejecting Gile's position, Nicholas denies that the existence of substance can be demonstrated. This leads him to limit severely, as does theory three, the manner in which we can talk about one object coming from another. In fact, about all that he is willing to admit in regard to change is that "nothing is generated without being preceded by something to which the emergent being had a natural ordering in its emerging," that is, "b" naturally follows "a".[57] Again, this follows closely theory three and its rejection of annihilation and creation *ex nihilo* in change. Nicholas does not, however, carry his assumptions to the logical conclusion set out in theory three, that corporeal things are substantially diverse. Instead, his rejection of hylomorphic composition leads him in an entirely different direction, toward accepting atomism as a means for accounting for change.[58] Since Henry is apparently unaware of this ramification of Nicholas's thought, it would appear that he has come across these ideas in an intermediary source that I have as yet been unable to identify.

As radical a departure as the third theory seems to be from the Peripatetic way of thinking, Henry does not dismiss it outright. Instead he concludes:

> certainly, this philosophy if it were frequently used and approved as custom over a long period of time, as has been the case with the Peripa-

tetic philosophy, would perhaps not seem so different from the Peripatetic philosophy.[59]

This statement seems to express, at the first glance, an openmindedness toward its acceptance. But the lack of customary use is precisely the point that allows Henry to sort through the morass of irrefutable opinions and theories just presented and arrive at an accepted solution.

The fact that the Peripatetic philosophy is accepted by Christians and that Peripatetic books are used in the *studia generalia* of the day are marks of divine favor (*esse signum specialis divinae dispositionis*).[60] This does not mean that the Peripatetic philosophy is without error. Just in case this is not obvious, Henry provides his students with a brief list of errors so that they are made aware of problems that may arise. If these errors are removed, what remains is a useful body of knowledge that provides rules for attaining knowledge (*regulas sciendi*) and weapons for confounding the errors of pagans (*erronea vel dubia gentilium philosophorum*).[61] So it is that the alternative theories do not have to be taken seriously and the Peripatetic matter-form theory, with Henry's Augustinian modifications, reigns supreme.

3. Angelology and Demonology

Having clarified *being*, the work of *in principio* can now be understood as the creation of matter and form. Both are created from nothing in the initial instant. As yet, however, they are not joined. The act of coming together, of form informing matter, involves motion, and *in principio* stands before all motion. Therefore, no thing as a hylomorphic entity, as a matter-form entity, yet exists. Still, there are beings who can properly be said to exist *in principio*. Between God, who is neither matter nor form, and creatures, who are both matter and form, there exists a third class of beings whose members are made solely of forms. This includes angels, intelligences (the movers of the planets), and demons. Since the members of this class do not require matter for their existence, they can properly be said to exist when their forms are created, and this was commonly thought to take place *in principio*.

Reference to angels is conspicuously absent from the Biblical account of creation. Nonetheless, there was really no doubt in the minds of most medieval commentators that Moses intended the terms *heavens*, *light*, and so forth to convey information about angels. As the most perfect of beings apart from God, there is really no other place to put their creation. Accordingly, most commentators followed Augustine's lead, even if they disagreed with his suggestions regarding instantaneous creation, and read into Genesis 1:1–2 an intended discussion of angels.[62] To quote the *Glossa ordinaria*, which in Henry's day was commonly attributed to Walafrid Strabo (d. 849): "Heavens [refer not so much to] the visible firmament as to the empereum, which is immediately filled with angels."[63]

Strabo's statement brings into focus the major questions concerning angelic being that have bearing on science and creation. Does *immediately* mean directly by God in the very first instant, or are angels created successively over time, one from another? How do angels fill the empereum? Do they have corporeal being, extended being, or some other type of being? Does the plural, *heavens* imply multiplicity in number, in species, or are angels all of the same species? Finally, if angels are created in the empereum, is that their only abode or do they operate in and have influence over the lower regions? These crucial questions concern an important part of creation. Angels come between men and God. Their perfect being provides an insight into the ways of creation that cannot be derived from creatures alone. Consequently, to ignore their existence would leave a part of creation, the part that lies closest to God, unexplored, something few medieval commentators were willing to do.

Henry notes that his own opinions about angels follow infallible teaching (*secundum illam infallibilem doctrinam*).[64] When he speaks of angels (also demons and celestial movers) he is discussing that class of beings known as rational creatures (*creaturae rationales*), that is incorporeal creatures whose primary function relates to their intellective capacities.[65] These rational creatures are created in the very beginning (*ab initio*) in a pure and undifferentiated state. Prior to their fall and their reception of a particular form of knowledge, they are void of all species and unique character (*nuda ab omne specie et habitu sicut tabula rasa*). They were created in this state and in their entirety *in principio* by God. Accordingly, angels do not come from one another but only from God.[66] Moreover, since all angels come directly from God, there are no evil angels in the beginning; all are good. To hold otherwise would go against the teachings of both philosophy and theology (*omni verae doctrinae philosophicae et theologicae repugnet*). Evil is an accidental quality (*omnis natura rationalis mala et perversa est solum accidentaliter talis*) that is acquired at the time that some angels fall, when they commit an act against God. Thus, the Manichee suggestion that there is a God of evil, or any other doctrine that comes close to this position, is strongly rejected by Henry, as it was by most Christians.[67]

Angels are located *in principio* in the empereum, the region of the heavens that lies beyond the sphere of the fixed stars.[68] Being located in this instance does not mean occupying space. Since angels are invisible and have no dimensions (*sit invisibilis et inextensus*), they cannot occupy space. Consequently, an infinite number could be packed into the space of the empereum (or on the head of a pin). In reality, however, there are not an infinite number of angels. Each is created within the confines of a strict hierarchy (*secundum gradum suae perfectionis*) as Dionysius (pseudo, fl. ca. 500) and others in the neo-Platonic tradition had amply explained.[69] By virtue of their position within this hierarchy, each is assigned a place. Therefore, when it is said that the heavens are full of angels, it is meant that each assigned place is full in much the same way that a city is said to be full when its houses are full.[70] Thus, angels are at least distinguished into degrees or grades. Beyond this basic conclusion, Henry doubts that much more can be said regarding numbers. He notes that some doctors attempt to argue that there is only one angel in each species (*solum unum individuum sub una specie*) even though the reasons given in support of this position may not be sufficient (*dicant rationem non habent sufficientem*).[71]

Henry's reluctance to pursue the problem of the number of angels is characteristic of his treatment of angelic being. Basically, he has very little interest in angels *per se.* Unlike his thirteenth-century counterparts who devoted more attention to angels than to the work of the six days, he comments on Strabo's statement with only a few brief remarks about place and number. In fact, most of the information just presented regarding angels is not set out within the context of the six days, but rather in conjunction with the fall of man, Genesis 3:1–7. Within the latter context, angels are important because it is from the goodness of angels that demons arise. So even here he is not so much interested in angels *qua* angels as he is in pursuing angelology in preparation for a more pressing science, demonology.

Henry's belief in demons is predicated upon two pieces of evidence. First of all, he is convinced that there are numerous phenomena that take place in the world of nature that cannot be accounted for in natural ways: revelation, inspiration, prophesy, magic, and so on. For Henry these are given facts of life which he cannot explain in the same way that he explains combustion, growth, and other natural phenomena. It is obvious, therefore, that intellectual beings must be active in the world. Secondly, many unnatural events are obviously of evil origin. The effects brought about by magicians and by those who commune with the dead (necromancers) are not within the domain of good angels. Thus, there are not only good but also bad and perverse rational creatures (*quaedam rationales subiectae undecumque malae et perversae*).[72] In case there is any doubt about the existence of the latter, one has only to turn to Henry's most important source in these matters, Augustine's *De civitate Dei*, to find long discussions regarding the dangers "of the illicit arts connected with demonology."[73]

Demons, according to Henry, are fallen angels. They arise when part of the society of angels goes against God's will and is expelled from the empereum. So there can be little doubt that demons, like angels, are rational creatures. They are not perverted men or men in different bodies. Demons are not made of skin and bones. Moreover, there can be little doubt that demons come from God; they are not the creation of some other god and do not exist *ab eterno*.[74] Once expelled from the empereum these fallen angels take up, or are assigned, residences in the region that lies below the empereum, and it is here that they work their evil ways.[75] Exactly how demons influence men and the elements is a subject to which Henry devoted a great deal of time. He seems, in later years especially, to have had a profound interest in the evil forces that were at work in the world around him. Unfortunately, a consideration of these evil forces lies outside the scope of the present study. Evil forces pervert nature and direct it from its normal course of events. This study seeks to discover how nature normally functions. Accordingly, demons and angels must be left at this point for the work that lies ahead, the creation of that part of being that lies below the empereum.

CHAPTER III

Day One: Light, Darkness, and Perspective

And God said, "let light be made," and light was made. And God saw that it was good. And he divided the light from darkness. And he called the light, day and the darkness, night. And the evening and morning were made, day one. Genesis 1:3–5.

In principio concerns the metaphysical. There is no other way to describe the initial work of creation. Such work takes place before things have form, before there is anything physical. All of this changes with God's fiat, "let it be made." At this point what was without form becomes formed and physical. Nature as we know it starts to take shape, although not yet in its final form or disposition. Final dispositions are added during the latter three days when God ornaments his work. The initial fiats only rough out the design: light from darkness, day from night, waters from waters, dry land from the seas. Later this design will be filled in. But we are not yet concerned with the filling in of design. First the plan must be laid bare as it was distinguished or set out by the Master Artisan during the first three days.

Laying bare the Creator's design brings science squarely into focus. The revealed word of itself cannot lead to a full understanding of nature. In this regard Henry would not agree with Luther, for example, when the latter discards the "needless opinions" of the philosophers and turns "to Moses as the better teacher."[1] Using science and the works of scientists to help interpret Scripture is not only desirable, it is necessary, as Roger Bacon (d. 1292), a century before, had steadfastly maintained.[2] Ignorance of the way things are leaves the theologian vulnerable to the attacks and ridicules of philosophers. If nature and Sacred Scripture cannot be reconciled, the way is left open for the rejection of the latter's teachings–even teachings that do not pertain strictly to nature. Therefore, "the perfect theologian ought to be not moderately learned in the human sciences" (*non mediocriter eruditum*).[3] In fact, Henry follows this suggestion of Augustine to the conclusion that knowledge of the human sciences is essential. No apology is needed for what follows, for unless we fully understand the workings of creation, the Creator too will fall beyond our intellectual and affective grasp.

The fact that the first manifestation of the divine fiat is light can only give to light a place of special importance in the universe. Surely, few medieval commentators went to the extremes of Platonists such as Robert Grosseteste (d. 1253) and assigned to light the role of first substantial form in the universe from which all subsequent forms in part derive,[4] but neither would most have denied that light occupies a very special place in nature. The reasons for this go well beyond its place of priority in creation. Light assumes many obvious and crucial functions in the world. Its presence and absence provides an instrument for time, for dividing day from night.[5] When it is present, it is also the instrument of vision, man's most noble sense. Through vision we have contact with not only objects in close proximity but even the remoteness of the heavens, which are themselves illuminated by light. Finally, since light originates in the heavens and is visible on earth, it is the one aspect of creation that is common to both the superior and inferior regions, acting as a bond to draw the two together. So close is light to God that Scripture speaks of the Word as "the light of men," the light that "shines in the darkness" (John 1:4–5). It is therefore not at all surprising to have an entire science devoted to light, the "notable and pleasing science called perspective" (*notabilis et delectabilis scientia quae perspectiva nominatur*).[6]

Although Henry has very little to say about the content of perspective in the *Lecturae*, he does praise the utility of this science highly and suggests that it helps in reaching an understanding of creation.[7] This is hardly unexpected, since he had earlier written an entire treatise on the subject, his *Quaestiones* on the *Perspectiva communis* of John Pecham (d. 1292). A few words about this science of perspective are in order, therefore, before turning, in section two of this chapter, to a description of the product of the initial fiat, light, and in section three, to the manner in which this fiat brought light into being. This will help to broaden our understanding of late medieval science as well as to provide the information that is needed for interpreting the work of the first day.

1. Perspective

Generally considered, perspective is the science of light. However, what sets perspective apart as a science is not so much the fact that it has light as its proper subject as the fact that it considers light in accordance with mathematical rules. The properties of light are close enough to the abstractions of mathematics to allow them to be equated to lines and geometric figures and to be treated in much the same way as the geometer treats abstract bodies. This makes perspective similar to, but more than, mathematics. As Pecham had earlier suggested:

> geometry considers the line as a mathematical entity; perspective considers the line more as a natural than a mathematical entity.[8]

But perspective is also not strictly speaking a physical science. Physics considers things *in materia*, as they exist in nature; perspective considers the same things *absolute*, without regard to particular circumstances.[9] Perspective thus lies between the physical and the mathematical, abstracting from experience the ideal case and describing that ideal case "with numerical calculations and geometrical demonstrations" (*arithmeticis calculationibus et geometricis demonstrationibus*).[10] The sophistication with which the medieval perspectivist was able to carry out this task is frequently quite great. Henry's own division of rays into three categories–direct, reflected, and refracted (*recta, reflexa, et refracta*)[11] –provides a convenient outline to illustrate this point, as exemplified in his *Quaestiones super perspectivam.*

Direct or *rectilinear* light is the light that radiates in straight lines from a luminous or light-producing source through an intervening medium to some object. How this takes place and the physical properties of light that pertain to radiation will be taken up later (sec. 2). What is of crucial importance to the perspectivist is how radiation can be described mathematically. It was commonly thought, (and Henry devotes an entire question to this issue) that when light leaves its source, it spreads out through the medium in pyramid-like fashion; that is, it is "multiplied pyramidally" (*multiplicatur per pyramides*).[12] This assumption leads to two further problems: whether the initial source from which the light radiates is a point, and whether the rays of light as they spread out are continuous or discontinuous. Henry resolves both problems in the *Quaestiones*, concluding that light does originate from a single point and that light rays spread as they radiate and are therefore continuous.[13]

Once the basic properties of direct rays are understood, this information can be applied to specific phenomena with an eye toward explaining how these phenomena can be accounted for by using the mathematical properties of light. Thus, one finds in treatises on perspective discussions of pinhole images (the appearance of a round image on a wall after solar light has passed through a triangular aperture), shadows (especially the shadows that appear during eclipses), the apparent magnification of luminous objects at a distance, and vision. Each of these phenomena can be explained, as Henry does in the *Quaestiones*, by applying mathematics to rectilinear propagation.[14]

Reflected radiation is radiation that is bounced or reflected at an angle from a mirror or some other hard surface. Mention of this type of radiation led perspectivists to consider the angles at which the rays are reflected, an activity that can become geometrically quite involved as various combinations of flat, curved, concave, and convex surfaces are imagined.[15] The applications of this type of radiation to natural phenomena are essentially twofold: first, the consideration of mirrors leads to the problem of image formation and the geometric computation of where images appear (Henry devoted one brief question to this problem in his *Quaestiones* on the *Perspectiva communis*)[16] ; and second, there is the rainbow, which Henry suggests can be accounted for by the light reflected from the curved surfaces of the individual raindrops that comprise a rainbow.[17] Reflection is also involved in accounting for the luminosity of the stellar bodies; however, this issue relates more to astronomy than mathematics and so will be discussed in a later chapter (chap. IV, sec. 1).

Refracted radiation involves the bending of light as it passes through media of varying densities. Henry devoted only one question to this problem in the *Quaestiones*, and his interest in this question clearly lies not with an abstract consideration of refraction, but rather with the observational difficulties that arise when light passes from the heavens through the atmosphere to an observer in the center of the universe.[18] Once again, the attention that he gives to perspective complements his interest in astronomy quite nicely. He notes, in concluding his discussion of the refraction of stellar light, that "astronomers ought to be experts in perspective" (*astronomi experti debent esse in perspectiva*) so they can determine to what extent the rays of light coming from the stars are bent by refraction.[19] This may in part explain what drew him to the study of perspective in the first place.

Light is not the only quality that radiates its influence over distance; heat, cold, and the occult qualities of stars also have effects on distant objects. Accordingly, it might be wondered whether perspective is limited strictly to the consideration of light. The general impression received from the writings of perspectivists is that it is. However, since Henry at one point seems to indicate that it is not, a few brief comments on this topic are in order.

In his questions on the *Perspectiva communis,* Henry pointedly argues that if "rays of hotness and coldness" (*radii caliditatis et frigiditatis*) are subject to laws, they do not follow the same laws as light since they do not radiate, reflect, and refract in the same way that light does. This leads to the conclusion that "visual perspective would not apply" (*non sufficeret*) to these other qualities.[20] In *De reductione*, which I have argued is a later work, this position is reversed. There the suggestion is made that some of the phenomena observed in nature might be explained "on the basis of the direct radiation and the reflection of the rays of the other sensible qualities with respect to subjects or objects."[21] The latter position is not expanded in *De reductione*, nor are the consequences of it pursued in any later works. Nonetheless, the suggestion that perspective may have broader applicability to other qualities does say something about his faith in the ability of the mathematico-physical sciences to account for the everyday events of nature.

Henry, like his contemporary Nicole Oresme (d. 1382), firmly believes that geometry provides a very useful tool for describing nature. Even more than Oresme, he is willing to carry this tool into the field, so to speak, and apply it to even the most ordinary experiences.[22] Therefore, the suggestion that perspective may apply to other qualities is not at all out of character with the general tenor of his thought nor with the general tenor of the science of his day.

Given Henry's deep respect for perspective, as evidenced by his *Quaestiones* and later statements in the *Lecturae*, and the general high regard most medieval scientists had for light, it might seem strange that so few of his contemporaries shared his enthusiasm and wrote treatises on perspective.[23] The lack of interest in mathematical perspective is not, however, really so surprising. Mathematical treatments of nature are not a salient feature of medieval science.[24] Astronomy, of course, presents a major exception to this rule, but in this case incentives were high. Even the general public could be aware of the obvious rhythmic patterns of the heavens and the need to calculate, however crudely, the motions of the stars and planets. Astronomers and astrologers have always had a ready market for their goods. Perspectivists were not so fortunate.

The connections between number and light are not readily apparent, nor are the consequences of these connections of much immediate benefit. Those phenomena that perspectivists handled with some sophistication—rainbows, pinhole images, vision, reflection—fall into the realm of common experiences whose mathematical clarification presents an interesting curiosity but little more. Moreover, the medieval scientist had other reasons for being interested in and studying light, and it is these that attracted most of their attention. As a participant in the Aristotelian matter-form world, light had to measure up to the general rules that govern the universe. Its very nature had to be explained; its relationship to other qualities such as color, clarified. Such are the concerns of the following section, concerns that were of almost universal interest throughout the Middle Ages.

2. The Physics of Light

When God created light, he created something that has the perplexing capacity to act over distances. A lighted or luminescent body seemingly of itself produces an effect that is manifest apart from that body: candles illuminate rooms; the light of the heavens shines upon the earth. In coming to understand light, it is this perplexing capacity, with all of its peculiar characteristics, that needs to be explained. What is a ray of light? How does it shine? What is darkness? What role do air and the other media through which light passes play in transmitting light? And perhaps most importantly, what is light?

In response to this last question, there should be little doubt, given the foundation established *in principio*, that Henry's answer will be given in terms of matter and form. This is not because other descriptions of light were unknown. Corpuscular theories of light had been advanced in antiquity and were well known throughout the Middle Ages. Aristotle, to mention one obvious source, discusses and rejects the notion that light is either fire or a body.[25] But if light is not a body, what is it? As its active characteristics indicate, it must be a form. Specifically, it is the form of a luminous body which, when that form is joined to a subject, expresses itself as the qualitative disposition "light" or "*lux*" (*lux signat qualitatem corporis luminosi prout est in corpore luminosa*).[26]

Whether light is an accidental or substantial form, Henry never really says. One school of thought on this subject maintained that light is the substantial form of luminous bodies; light is the form that makes the sun the sun.[27] Aquinas and others rejected this definition in favor of the position that light is an accidental quality. According to Aquinas, "just as heat is an active quality deriving from the substantial form of fire, so light is an active quality deriving from the substantial form of the sun, or any other body that is self-illuminating."[28] This latter definition is most consistent with the general tenor of Henry's discussion of light.

Light, as a qualitative disposition, can be understood in two ways; as a primary or as a secondary quality. In a psychological sense, light is a primary quality since it is perceived directly of itself (as color) and does not require more fundamental qualities for its action. This contrasts to secondary qualities (*intentiones visibiles secundariae:* shape, size, number, and so forth), which are perceived through other qualities–light and color in this instance–and not through their own qualitative forms. Light possesses the capacity to act in and of itself and is not dependent, psychologically speaking, on other qualities.[29]

In an ontological sense, light is dependent on other qualities and, therefore, is a secondary quality. This second sense is briefly alluded to in the *Quaestiones*, where Henry relates light to fire, arguing that "fire is most productive of that secondary quality, namely light" (*ignis maxime productivus est illius qualitatis secundariae, scilicet lucis*).[30] This suggestion is expanded at some length in *De reductione*, where a much more detailed discussion of primary and secondary qualities is presented. In this work Henry explains that light derives from four primary qualities (hot, cold, wet, and dry), thus helping to explain its heating and drying actions.[31]

Of the two senses in which light can be understood to be a quality, it is the first, its psychological sense, that is of immediate concern. The second sense pertains more to the role light plays in generation and corruption, a topic that will be taken up later (chap. VI, sec. 1).

Having resolved what sort of quality light is, the next problem to be considered is how this quality acts over distances. As a quality, light can act only in conjunction with potency. Initially it resides in or informs the potency of its subject, a luminous body. Thereafter, in order to radiate, it needs a potency that will allow it to act apart from the potency of its subject. This additional potency is supplied by the medium that comes between a luminous body and what is illuminated, between a candle and a wall, for example. The medium–air in this case–receives the forms of light and transmits them to the object. Radiation thus is the process by which the forms of light informing a luminous subject move–the medieval scientist would say the forms "are multiplied"–from that subject through the medium to a distant object. The forms that move through the medium are not, however, the same as the forms of the luminous subject. If they were, the medium would become "illuminated" (informed by light and thus luminous) and a second candle someplace in the room, upon receiving the forms from the first candle, would become "lighted", which it does not. Accordingly, it is obvious that the forms moving out through the medium are not the forms of light (*lux*) but only the likenesses of light (*lumen*), which convey the activity of light (*lux*) to distant points. Light (*lux*) multiplies by sending likenesses of itself (*lumen*) through the intervening potency or medium.[32]

The capacity that light has to multiply itself through a medium sharply sets it apart from color. A simple test from experience makes this clear. If a light is introduced into a dark room, we see it; if a colored object is brought into the same room without light, we do not. Color and light are both accidental qualities, but only light multiplies *per se*.

How then is color multiplied? It is multiplied with the aid of light. When light traverses or is multiplied through the medium, it changes that medium from its state of potential transparency to a state of actual transparency (*lumen est actus dyaphani secundum quod dyaphanum [est] . . . , intelligit pro dyaphanum, corpus transparens*). In the process, it removes the barrier–a lack of transparency or what we would refer to as darkness–that prevents color from multiplying its species, thereby allowing it to become visible.[33] Once that barrier is removed, color and light act in much the same way; that is, multiplying by likeness, travelling in straight lines, reflecting and refracting, and so forth.

Both light and color are what Henry refers to as spiritual, as opposed to material (hot and cold), qualities (*qualitas est duplex . . . corporalis et materialis, . . . spiritualis et immaterialis*): they pass through the medium without effecting it; they have no contraries; they do not physically alter the object they inform.[34] But insofar as their ability to act is concerned, they are sharply distinguished. Only light possesses the capacity to render the medium transparent, to change darkness to light.[35]

The multiplication of both light and color through a medium involves motion, the advance of a form from one point in space to another. Motion normally implies time; according to Aristotelian physics, motion cannot take place instantaneously.[36] But what about light? Does its multiplication require time, or is light an exception to the rule? Henry takes up this problem in both *Notata de anima* and *Quaestiones super perspectivam*, each time resolving the issue in a slightly different way. The key to understanding the consistency that lies behind the two discussions is his definition of the term "very rapidly" (*subito*).

In the questions on the *Perspectiva communis* he defines *subito* as meaning "in an instant" (*in instanti*) and "successive" (*pars post partem*, one part after the other). Given this definition, he goes on to argue that light "very rapidly illuminates the entire medium, both extensively and intensively" (*subito totum medium extensive et intensive illuminatur*).[37] In other words, both with regard to extension and intensity, light is multiplied through a medium seemingly in an instant, but not instantaneously or without time. Multiplication involves successive motion, and successive motion takes time.

That multiplication *in instanti* is not instantaneous is made clear in *Notata de anima*, where Henry defines *subito* more narrowly as "nonsuccessive" (*non pars post partem*).[38] Assuming this latter definition of *subito*, he goes on to reject instantaneous multiplication. The reasons for this are obvious. First of all, instantaneous multiplication would imply that the medium offers no resistance, which in turn would lead to the assumption that rare and dense media would multiply light with equal speeds, an assumption that Henry argues is false. Likewise, for a power to multiply something with infinite speed, it must have infinite power (*esset infinitum virtutis*), which is also impossible.[39] "From which it follows that all illumination, in so far as it is from its part, is [not] produced instantaneously but successively" (*quantum est ex parte sui [non] fit subito sed successive*).[40] He is therefore willing to grant that light multiplies very rapidly, seemingly in an instant, but not so rapidly that one part does not follow another. The multiplication of light requires time.

The discussion of instantaneous multiplication in the *Notata* is concluded with mention of one possible exception, that is, the multiplication of intensity. Henry notes that it is possible that even though light may not initially be multiplied instantaneously, once light is present, changes in intensity may occur instantaneously (*licet lumen non generatur subito secundum quantitatem in spacio, generatur tamen subito secundum intensionem in formis*).[41] This raises an interesting aspect of multiplication that needs further clarification. We have already seen how light multiplies "according to quantity in space." What now needs to be discussed is how once light is multiplied in space, its intensity can change. This necessitates the introduction of a concept that was frequently utilized in the discussion of motion and that Henry brings to bear upon his discussion of light, the concept of the intension and remission of forms.[42]

What is meant by the intension and remission of forms is simply graded rather than abrupt change from one form to the next. When an object becomes lighted, it can become lighted in one of two ways: either it can go immediately from having no form of light to having the form of brightness that ultimately characterizes the lighted object; or it can become lighted by going from its unlighted state through increasing degrees of brightness to its final form of brightness. (An example of the difference involved, however imprecise, would be the contrast between turning a light on with a normal light switch versus using a rheostat or dimmer switch. The latter change obviously takes place by degrees; the light is *intensified* over a period of time. The former seemingly takes place without degrees, although it could pass through degrees of brightness at a rate that is simply too rapid to observe.) Henry applies this doctrine to changes in the intensity of light.

The observed brightness or intensity of a light, according to Henry, can change in one of two ways: either the light itself (*lux*) can shine more or less brightly, be intensified or remitted; or its likenesses (*lumen*) can be effected by the multiplication process, causing them to decrease in intensity or to be remitted. Regarding the first way, since light (*lux*) originates from the individual points of light that make up a luminous body, when we seek to understand how a luminous body can shine more brightly (be intensified), what is being sought is some means to explain how points of light can shine more brightly. Henry argues that this can happen in one of two ways: either a point of light can have degrees of light added to it (*per novi gradus luminis additionem*); or it can become more closely packed together with adjacent points of light (condensed, *per eius subiecti condensationem*). In either case, the quality of light will be intensified.[43] It is important to note in this regard that a single point of light cannot be intensified *per se*, cannot be a self-moving power, nor can it be intensified by points of light adjacent to it. If it could be, then theoretically the sun's intensity would change over time (it would become brighter), which it does not, "even in one thousand years."[44] Therefore, light (*lux*) is intensified and remitted only when some power outside the luminous body adds or subtracts degrees of light, or when the volume occupied by a given number of points of light changes.

The second way that the intensity of light changes is over distance, as its likenesses (*lumen*) are multiplied through the medium as rays of light. In this regard it was usually assumed (as experience easily confirms) that light weakens over distance.[45] Henry agrees with this assumption and goes on to describe the remitting process in terms that were commonly used in conjunction with the doctrine of the intension and remission of forms. If light did not decrease in intensity as it multiplies over distance, then we would say that it is composed of uniformly uniform rays (*radius uniformiter uniformis*). That is to say, the rays not only would not change in intensity over distance (being uniform), but they would also be consistently the same over their entire distance. This clearly is not what happens when light multiplies, because if it were, then a single point of light would have infinite power and could illuminate the entire universe.[46]

The rays of light are therefore not uniform but difform (*difformis*); they change in intensity as they multiply. This change can take place in one of two ways: either the remission of forms (decrease in intensity) can be at a uniform rate (*uniformiter difformis*), or at a nonuniform rate (*difformiter difformis*). Henry accepts the latter case, arguing that if only a single factor or weakening agent were involved, then the remission of light over distance would be uniform (uniformly difform). However, since in most cases several factors are involved, he suspects that light is not uniformly remitted and that light rays are difformly difform. Ultimately, through this process, all of the intensity of the ray will be remitted as it reaches the point of zero or no degree of intensity—that is, darkness (*versus non gradum*).[47]

This is *how* light is remitted over distance; we still need to understand *why*. Why do light rays tend toward zero degree of intensity? Two reasons can be given for this: either the process of overcoming the resistance of the medium (*ex parte resistentiae*) requires effort, which in turn weakens the power of the ray; or the prospect of extending an infinite distance (*ex parte distantiae*) is so repugnant to the normal course of nature that the light ray would of itself tend toward total remission rather than extend an infinite distance. Curiously, Henry favors the second explanation, arguing that even if God were to destroy all resistance (*si Deus suspenderet omnem potentiam resistitivam medii*), a ray of light would still tend toward zero intensity *ex instinctu naturae*, from something placed in it in the course of nature.[48]

The reason for characterizing this position as a curious one is that in an earlier discussion of pinhole images, Henry had rejected an *ex instinctu naturae* explanation in favor of a rather complex and apparently unprecedented geometrical explanation.[49] And yet here, where he has a perfectly obvious reason for rejecting infinite multiplication on the basis of the resistance of the medium, he ignores this explanation in favor of a built-in controlling mechanism in nature. However, the difference between the two cases is significant. In the case of pinhole images, what was possibly accounted for by nature was a tendency of light to go toward circularity. This would have easily accounted for pinhole images, but it would also have violated an even more fundamental law of nature, rectilinear propagation. In the case of remitting intensity, it is the fundamental law of nature that is being accounted for by nature, not the anomaly.[50] Clearly, nature can only have a tendency toward what is its normal course and not some other course.

I have dwelled on the doctrine of the intension and remission of forms at some length because it is applicable to many aspects of Henry's thought. Not only light, but potentially any quality, can be intensified and remitted in much the same way. *Intentio* and *remissio* are common terms in his vocabulary, so common that he seldom stops to explain what they mean. Then too, discussing intension and remission in conjunction with light helps to reinforce the fact that this doctrine is firmly grounded in experience. Prior to electric light switches, most changes in brightness that would normally be experienced would be graded ones, and the same would be true of changes involving heat, cold, motion, and the other qualities that medieval scientists sought to describe using this doctrine. The doctrine of intension and remission accounts for the way things are. It focuses squarely on the world as perceived and attempts to account for nature on its own terms. If, in the end, visualizing the world in these terms fails as a system for scientific explanation, it is not because this system is too detached from nature, but because it is too close to it. Medieval scientists discuss the world that immediately surrounds them. It is only the science of later periods that will gradually withdraw from and idealize nature in order to wrest from it the abstract laws that are apparently better suited to account for its actions.

Understanding intension and remission fills in the last details needed to deal with the medieval conception of light. *Lux* is the initial active form that is light. *Lumen* (as either a likeness or mathematical ray) provides the mechanism for securing action at a distance. The transparency (the medium) adds the potentiality that allows that mechanism to act. Intension and remission bring in the element of variability that accounts for the subtle changes that are obviously present in nature. Each ingredient, each aspect of light, can of course be further clarified. The reliability of vision needs to be accounted for. Anomalies, such as rays being perceptible in some media and not in others, need to be explained. Opacity as an unchanging property of some bodies must be brought into the orderly balance of nature.

These, plus a host of similar problems, both geometrical and physical, make up the content of the medieval science of light. Sorting through the many prior opinions, the related common experiences, the applicable rules of nature, and providing a reasonably complete description of light, comprises one of the primary activities of the medieval scientist. Such were the ends that Henry attempted to achieve in his questions on the *Perspectiva communis* and in related sections of his commentary on Aristotle's *De anima*. Having reached this "scientific" understanding of light, the task left for the remainder of this chapter becomes immeasurably simpler, that is understanding how and when God created light.

3. Darkness and the Creation of Light

Deciding when light was created has always been a source of some difficulty for commentators on Genesis. The apparent inconsistency between having the sun, the source of all light, created on the fourth day, while light itself was created on the first day, is so significant that it has led to numerous speculations as to what precisely "*fiat lux*" means. In resolving this inconsistency, Augustine set the tone for much of the discussion that was to follow. *Lux* can refer to corporeal light, which illuminates the world. It can also refer to spiritual light, which in Augustine's view is the same as referring to spiritual creatures or angels. It is this second interpretation that Augustine preferred, arguing that "heaven and earth" refer to "spiritual and corporeal," thus making "*fiat lux*" a further statement about spiritual beings.[51]

It is also the spiritual interpretation of light that Henry is following when he argues that the separation of light from darkness refers to the separation of good angels from bad angels.[52] Henry, like most commentators, was willing to interpret light as a spiritual entity, thereby lending further credence to the belief that angels were created *in principio* (chap. II, sec. 3). But this view does not deny a corporeal interpretation of *fiat lux*, an interpretation that was taken up and discussed by Basil (d. 379), the Venerable Bede (d. ca. 672), Strabo, and others.[53] In this case, the apparent inconsistency had to be resolved.

Determining how and when the first light was created begins with *in principio*. Even before God created light, "darkness was upon the face of the abyss." But does this mean that darkness precedes light, or were light and darkness somehow present together from the very beginning? Many commentators simply assumed that darkness came first, arguing, as did Basil, that light "dissipates the darkness and illuminates the world."[54] Augustine, on the other hand, perhaps because of his greater sensitivity to the dangers inherent in assigning priority to darkness, made light and darkness coterminous in the initial instant in which all things were created.[55]

Henry sides with the latter opinion. He agrees with Augustine's assumption that light was present *in principio,* and that this does not exclude the possibility of darkness also being present. Light as *lux* has no effect. It is only as *lux* radiates *lumen* that illumination takes place. But the multiplication of *lumen,* as we have already seen, is a successive process, one that requires time. *In principio* there was no time. Therefore, even though *lux* was present *in principio*, it had yet to multiply *lumen*, and so there was darkness over the face of the abyss. *Fiat lux* means *fiat lumen* (*esset dicere fiat lumen*). This, Henry concludes, "seems sufficiently probable and more truly saves the words of Scripture" (*videre satis probabilis et quod salvat realius verba Scripturae*).[56]

But if *fiat lux* means *fiat lumen*, from what *lux* did the *lumen* radiate? Given the fact that the sun had not yet been created, there were two ways to dispose of this question. One way, which followed the opinion of pseudo-Dionysius, was to assume that the first light was the light of the as yet unformed sun (*lux fuit lux solis a principio . . . informis et incompleta*). This light remained in a pure and unformed state until the fourth day, when it became formed and imperfect (*sol habuerit tunc lumen quasi turbidum seu imperfectum*).[57] A second way to dispose of this question, one that Henry only notes in passing but which had considerable popularity because it was mentioned by Peter Lombard in the *Sentences,* assumed that there existed on the first day a small luminous cloud (*quaedam nubecula rotunda imperfecte lucida*) that performed the functions of the sun prior to the latter's formation.[58] Exactly how this cloud, or the imperfect sun, functioned is never made abundantly clear. What is important to note is that ways could be found to account for light even before the sun appeared. "God said, 'let light be made,' and light was made!"

The obvious consequence of bringing the first light into being is day and night. This too led to problems. How can day and night be accounted for with light and nothing else? (The heavens and earth have yet to be distinguished.) One school, which Bonaventure earlier had suggested was that of the Greeks rather than the Latins, maintained that light originally came into the world in an ebb- and flow-like manner. Day was made when light flowed into the world, night, when the light was drawn back (*ad tempus unius diei artificialis radiare et post suspendere*).[59] The more common opinion of the Latins was that the first light, when it came into being, had diurnal or twenty-four-hour rotation; it moved around the universe in twenty-four hours, just as the sun will when it comes into being three days hence (*illam lucem motam fuisse circulariter circa centrum mundi*).[60] However, rotation alone does not produce day and night. Night is the absence of light, so something must prevent light from radiating throughout the universe. Once again, this can be accounted for by considering how light is multiplied. Light requires the form of transparency for its multiplication. During the first day the earth was as yet unformed; it had not even received the form of transparency (*forma terrae non determinet sibi naturaliter dyaphaneitatem*). Therefore, it acted like a dense body in the center and cast the shadow of night.[61]

Fully describing the events of the first day led to other problems that may not be apparent to the modern mind whose historical senses have been dulled by evolution and the rapid recession of *in principio* to a dim and distant past. However, if *in principio* is relegated to a point in time only five or six thousand years before Christ,[62] then day one takes on definite attributes that can be pinned down with some precision.

The first light must appear at some place in the as yet unformed heavens; but where? Overhead, in the position of high noon, so that its rising and setting points can then be set out on either side? Henry's reply to this much-debated question is that the sun was originally created in the East, because wherever it was created, that place will become the point of sunrise (*oriens*) and the opposite point, the point of sunset (*occidens*). He therefore assumes that the first day begins at sunrise.[63]

Then too, the first day must fall at some time within the first year; but when? Since green plants will shortly (on the third day) spring forth from the earth, it would seem as though the earth were made in the spring, when plants begin to grow. After all, March is designated as the first month in the Bible (Exodus 12:2). And yet it might be argued that the earth was created in the fall when it was full of seeds (*in mense Septembris quando fructus in arboribus sunt maturi habentes semina in seipsis*), full of the potential that could later spring forth from it.[64] Which of the two positions was true? The students listening to the *Lecturae* were apparently left to decide for themselves, no doubt conjuring up in the process images of the balmy breezes of a warm spring day or the crisp, clear atmosphere of harvest times.

The nonmedieval mind might rebel at this point and wonder, so what? What difference does it really make? Light is light, darkness is shadow, and so on. All things came into being in the beginning. What use is there in pursuing these specific and unconfirmable details about seasons, horizons, and small luminous clouds? There is a very important use for such speculations and it goes well beyond a strict and literal interpretation of the Bible. Studying creation does lead to information about how God created the universe; this is true. Studying creation leads to an appreciation of the Creator; this is also true. But a use of equal importance is that studying creation provides an opportunity to reflect on standard and accepted scientific explanations in order to determine whether or not these explanations are adequate. In this sense, the commentary on the six days of creation provides more than simply an excuse for doing science; it becomes a *way* of doing science. It furnishes one means through which hypotheses can be universalized and extended to the beginnings of time and the outermost fringes of the universe. It focuses attention on problems that might otherwise not have been thought of. It uncovers anomalies and inconsistencies in nature that later scientific societies will derive from laboratory experiments. Commenting on Genesis and on the minute details that go into bringing a universe into being is, in short, part of the laboratory procedure of medieval science, a part that is totally consistent with a society that conducted most of its science within the confines of the lecture hall. That this is true will become even more apparent as we turn to what is perhaps Henry's favorite aspect of creation, the heavens.

CHAPTER IV

Day Two: The Heavens and Astronomy

And God said, "let a firmament be made in the middle of the waters." And he divided the waters from the waters, and God made the firmament. And he divided the waters that were below the firmament from the ones that were above the firmament. And it was made in this way. And God called the firmament, heaven. And the evening and morning were made, day two. Genesis 1:6–8.

Dividing the waters and establishing the firmament are actions that pertain to the heavens. There was little doubt among commentators on Genesis that "let the firmament be made" (*fiat firmamentum*) meant let the incorruptible orbs of the heavens be distinguished (*facta sunt et disposita . . . quantum ad omnes orbes*).[1] Bringing form to the heavens focuses attention on one of the most intriguing and yet puzzling parts of creation. The heavens are filled with events and entities that invite speculation without providing a great deal of information that can be used to reach firm conclusions about them. The stars that twinkle overhead yield scant information as to their make-up. The seemingly regular movements of the planets provide ideal subjects for mathematical computations even though the physical mechanisms that produce their movements remain obscure. Even the exact positions of the celestial orbs, as they fill the space that extends from the elements to the empereum, is something that medieval scientists could not describe with any degree of certainty and precision. Still, they tried, in this area perhaps more than any other, which prompted the numerous scientific speculations about celestial events and inhabitants that pertain to the second day.

The work of distinguishing the heavens on the second day is, in many ways, very difficult to separate from the work of the fourth day. The work of both days pertains properly to the heavens, while the acts of distinguishing the orbs (day two) and ornamenting them with their stars (day four) lead to many of the same scientific questions. Accordingly, there is some duplication in the material covered in the *Lecturae* pertaining to the second and fourth days. In order to avoid this duplication, I have made explicit a distinction that is implicitly, but by no means rigidly, adhered to in the *Lecturae*. Day two pertains most directly to the heavens as considered in and of themselves: the number of orbs, their motions, their physical make-up, and so forth. Day four pertains more to the heavens as they are connected to events in the lower regions, as they participate in what Henry calls the "golden chain" that ties men to the stars, and the events in this world to celestial influences.[2] In maintaining this distinction, some of the subtleties associated with the creative act must be ignored. I will, for instance, refer to the stars in this chapter as though they were fully formed, even though their final form was not added until the fourth day. Hopefully by sacrificing exegetic accuracy, a clearer understanding of the science that underlies this exegesis will emerge.

Separating matters relating to the heavens into two specialized areas is entirely consistent with Henry's thought. Heavenly events properly pertain to the subject matter of astronomy. However, astronomers consider these events in a number of distinct ways. This is clearly explained in what is possibly an early, elementary treatise on astronomy by Henry, the *Expositio terminorum astronomiae*.[3] Generally defined, astronomy

> considers the magnitudes of heavenly bodies; their distances above their rising and setting points; their swiftness, slowness, and other accidents of motion; and [related] natural events in these lower regions, namely: tempests, calms, sterility, fertility, and other things of this type.[4]

As such, astronomy divides into three (or two) parts: mathematical, natural, and (or) judicial astronomy. Mathematical astronomy (*astronomia mathematica*) considers quantities, distances, and motions by means of demonstrative mathematical methods. It endeavors to calculate precisely the paths that the celestial bodies follow (*non solum sermone sed etiam certa demonstratione patefacit*).[5] Natural astronomy (*astronomia naturalis*) attempts to uncover in the stars the causes of natural events (*ex virtute stellarum satagit reddere causas eventuum naturalium in hoc mundo*).[6] Judicial astronomy (*astronomia judicialis*) also deals with the natural events, hence the twofold division of astronomy into mathematical and natural. However, it goes beyond natural astronomy, narrowly considered, and attempts to make judgments about hidden things (*de occultis*) and future events.[7] Consequently, the distinction that is being utilized in the discussion of the second and fourth days parallels the distinction between mathematical astronomy and natural astronomy. Chapter six will deal with the latter, bringing together in the process two of Henry's favorite topics, the investigation of cause and effect (natural astronomy) and a critical evaluation of the merits of astrology (judicial astronomy). This chapter deals with mathematical astronomy (sec. 3) and two closely related topics, the physical make-up of the heavens and heavenly bodies (sec. 1), and the number and order of the celestial orbs (sec. 2).

tance of the ether by scientific thinkers of the sixteenth and seventeenth centuries, has often led to the conclusion that there was no serious challenge to the quintessence prior to the time of modern science.[15] Such an assumption is simply not true.

Several serious problems arose for both scientists and metaphysicians when the matter of the heavens and the elements were distinguished. Hence, a number of influential thinkers chose to reject the notion of an ether. Among earlier writers, Ambrose argues that the addition of a dissimilar part to creation would produce imperfection rather than perfection. Heavens and earth were created together and are one; one part does not have priority and superiority over the other.[16] A more closely reasoned argument given by Henry, William of Ockham (d. 1347), and others in the fourteenth century objects that if the matter of the heavens has potency for only a single form, then either the celestial bodies are all composed of the same form or there are as many types of matter in the heavens as celestial bodies. That is to say, if the quintessence is receptive of only one type of form, then either the sun and moon must be of the same form or there must be one type of quintessence for the sun, another for the moon, and so forth. Ockham found neither alternative acceptable. Henry admitted that the latter alternative was the more probable, although he ultimately concurs with Ockham and rejects the ether as a component of celestial bodies in favor of yet another explanation of the incorruptibility of the heavens.[17]

The explanation of celestial composition that Henry accepts accounts for incorruptibility in terms of form. Having rejected the need for a fifth element and accepted the notion that the heavens have matter-form composition, there was really no other choice. Henry suggests that through a vague and only briefly alluded to process, forms found in the inferior regions of the universe are somehow ennobled as they move toward the celestial region. When this process is completed—that is, when the celestial forms reach the heavens—they become incorruptible by virtue of the fact that they no longer have any contrary qualities associated with them (*annexae contrariae qualitates*). Therefore, the matter-form composits that make up the heavens could change because their matter is similar to earthly matter and hence has potency for other forms; however, they do not change because there are no contrary qualities present to produce change. Even though the matter that underlies the universe is one (*eiusdem rationis*), heavenly bodies are incorruptible because their forms are incorruptible (*formae naturaliter incorruptibiles*).[18]

What Henry has in mind with this explanation becomes clear when he discusses, in very physical terms, the composition of the stars. Numerous opinions had been put forth on this subject. It had variously been maintained that stars are: (1) gods, (2) flames, (3) a mixture of the four elements, (4) a congealed part of an orb, (5) a luminous material that is not part of an orb, (6) reflected solar rays, (7) the light of the empereum shining through holes or transparencies in the sphere of the fixed stars, (8) watery crystals, (9) fumes and exhalations of a more noble sort than those found in comets, and (10) a fifth element.[19] From this list, Henry chooses the third explanation; stars are composed of elemental mixtures that are transported to the heavens and placed in their respective orbs.

His reasons for choosing this explanation are various. First of all, stars are located in one portion of their orb; they do not occupy the entire orb. As such, they could have two possible origins; they could either be (a) condensed from the matter of their orb, or (b) made of an entirely different matter and then placed in their orb. The former explanation leads to difficulties. Stars and orbs do not have the same properties. Stars shine and reflect light, orbs do not. Orbs are transparent and invisible, stars are not. Therefore, it would seem unlikely that one could be condensed from the other. Furthermore, if stars were condensed from their orbs, the condensing process would cause the orbs to shrink or become rarified (*rariores fieri*), presumedly making the universe less full than it should be. And finally, once the condensing process begins, there is no reason for it to stop before the entire orb becomes a luminous star, which clearly does not happen.[20] Accordingly, Henry rejects the notion that stars are condensed from their orbs and assumes instead that they are composed of an elemental mixture that is first created in the region of the elements and then drawn together in the region of the firmament (*facta primo in regione elementi et post in regione firmamenti collocata*).[21]

Shifting the origin of the stars from the celestial orbs to the elements leads to other problems. If the stars are condensed from the elements, then what remains of the elements becomes rarified, thus raising an objection similar to that involved in condensing a star from its orb. In this case, however, Henry is willing to grant that rarifaction does take place, suggesting that initially the elements were more dense than they are now (*in maiore densitate creata quam modo sint*). Besides, it does not require a large quantity of elemental material to make bodies such as the stars, as experience makes clear. The amount of ash, and therefore the total elemental component, left when a log burns or a body decays in a tomb (*a corpore hominis consumpto in sepulchro clauso*) occupies considerably less space than the log or body did originally.[22] Thus, only a small amount of elemental material needs to be drawn from the elements and placed in the heavens to produce stars.

Transporting elemental materials into the heavens presents another problem. Although Henry never fully describes the physical composition of the stellar orbs, it is clear that they are, in his opinion, physical entities. They are formed on the second day in much the same way as water or lead congeals; they become firmly set (hence the designation *firmament*) as a series of concentric shells or spheres that stretch from the region of the moon to the region of fixed stars.[23] They are clear, firm, impenetrable, and have thickness. They move in a circular fashion, carrying with them their stars, once the stars are formed on the fourth day.[24] In fact, the image of glass globes spinning on fixed axes around the central earth, so commonly used to describe the medieval conception of the celestial orbs, seems to fit quite nicely the discussion in the *Lecturae*.[25] Consequently, they provide a barrier to the movement of the elements from the inferior to the heavenly region.

Since the orbs are fixed in place on the second day, God has to transport the elemental mixtures through these orbs on the fourth day in order to get them into place. However, moving such mixtures through the orbs would sever them (*fuisset penetratio vel scissio*) and cause them to break.[26] This problem proved to be so serious and problematic that Henry had to resort to supernatural causation to find a solution: "Just as the body of Christ crossed through closed doors and the heavens without breaking them, so too God could collect the stars, which were made below, in the most removed location of the firmament without such an inconvenience."[27] Therefore, the creation of stars is ultimately a miraculous event that falls outside the normal course of nature.

To summarize, there can be very little doubt from this discussion of the physical composition and origin of the stars that Henry considers celestial incorruptibility in very physical terms. Stars are composed of matter and form. Their matter is the same as the matter than underlies the elements; it is *prima materia*. Their forms are the same as the forms of the elements, with one exception. In being transported into the heavens, the stellar forms are stripped of contrary qualities and hence become incorruptible. Other than this, the inhabitants of the heavenly and earthly regions differ very little. They are all composed of mixtures of the elements. The exact nature of the mixtures of elements placed in the celestial orbs is not made clear in the *Lecturae*. About all that can be concluded is that heavenly bodies are not pure fire, as the Platonists had maintained. If they were, they would be self-luminous. According to Henry, however, the only self-luminous body in the universe is the sun. All of the other stars, including both the planets and the fixed stars, receive their light from it.[28]

How the sun illuminates the remaining celestial bodies led to further difficulties. In the case of the moon, for example, perspectivists suggested that the moon acts like a mirror, reflecting light from the sun but having no light of its own. Among other objections to this position, Henry notes that if the moon were a mirror, the spots on the surface of the moon would be actually spots on the surface of the sun (he assumes, without saying so, that this is not possible), and that we should see the planets and bright stars reflected in the dark portion of the moon, which we do not. Moreover, if the moon has no light of its own, there would be no way to account for the turbid light that remains associated with it during eclipses.[29] A second school of thought believed that the light of the sun either causes the moon to give off light itself or that the light of the sun is captured by the moon and then reradiated (*illuminatio lunae fit per incorporationem luminis solis in profundum corporis lunae*). This would help to explain why the moon remains lighted during eclipses, but not why the dark side of the moon normally has no light. If the moon is physically consistent throughout (*homogenius*), then the light of the sun ought to cause its entire surface to be luminous and not just a part.[30] Thus, there are objections to both explanations of how the moon receives its light, causing Henry to leave this problem unresolved.

A similar lack of agreement arose with regard to the light of the fixed stars. Unlike the planets, the light of the fixed stars varies; the stars twinkle. The perspectivists argued that this variation is caused by changes in the angles at which the sun's light reflects from the stars, changes that are brought about by the motions of both bodies. Henry rejects this explanation since it would apply to the planets as well as to the stars, and the planets do not twinkle. Pecham had anticipated this objection and went on to suggest that the relatively greater distance of the stars would cause them to twinkle, while the planets, which are closer, would not. However, even this clarification did not satisfy Henry; Saturn is almost as distant as the stars and it does not twinkle. So again he was led to leave the initial problem unresolved, arguing in this case simply that the stars twinkle because God made them that way.[31]

Resorting to the ways of the Creator to explain events in nature may seem, at first, unscientific. It is important to note, however, that in this case and in the other example of God's direct intervention in nature mentioned above, Henry is not resorting to miracle so much as to mystery. There are some aspects of nature that he simply cannot explain. And like Augustine, he is hesitant to make suggestions when evidence is slim and authority divided.[32] Accordingly, he simply notes that what exists, exists because God made it that way. He does not explain further the nature of its cause. Such scepticism is hardly unwarranted in the case of physical astronomy. When it comes to the physical make-up of the heavens, data was slim and authority divided. Such will also be true of the next topic to be considered, the actual number of celestial inhabitants and their placement.

2. The Number and Order of the Celestial Inhabitants

Since the medieval astronomer generally considered the heavens to be incorruptible, it may seem strange to raise the issue of the number of celestial inhabitants. However, the incorruptibility of the heavens does not rigidly rule out the possible appearance of new stars, nor does it guarantee that the number of stars that are known include all that exist. There are, as al-Battani (9th Cent.) had reported, stars around the South Pole that no one has even seen. New stars appeared in the heavens at the birth of Christ. Some in Henry's day maintained that comets, which come and go, are stars. And finally, there is an argument that may be uniquely Henry's: since Mercury only rarely comes forth from the sun's rays, there is no reason why there cannot be other planets that always remain within the rays of the sun and hence are never visible (*aliquae stellae erraticae quae semper in radiis solaribus, nobis invisibiliter versantur*).[33] In short, there are numerous quirks in the incorruptibility of the heavens that need to be explained. The advocates of incorruptibility had to proceed with caution when defending their position. For Henry's part, he found it more convenient to suggest that no one is certain about the number of stars except "he who created them and numbered their multitude" (*ille qui stellas creavit et qui earum multitudinem numeravit*).[34] Even so, he usually accepts as a working hypothesis that there are only seven planets and that the number of fixed stars remains unchanged. This leads to the obvious problem of how the stellar orbs are arranged.

The only firm evidence that we have for ordering the planets, according to Henry, is the fact that the moon is the lowest of the heavenly bodies, since it eclipses the sun and the rest of the wandering stars.[35] Beyond this, inferences can be drawn from the speeds at which the planets move about the earth (*persuasiones . . . ex tarditate motus et huiusmodi*), inferences which cause Henry to posit that Saturn is the highest planet with Jupiter directly below it. Other than this, little data is available to help position the planets.[36] Accordingly, he is willing to tamper with the standard "Ptolemaic" ordering of the planets–the moon, Mercury, Venus, the sun, Mars, Jupiter, and Saturn–as when he attempts to account for a change in the brightness of Venus reported by Varro (through Augustine).[37] If the sun is placed between Mercury and the moon, then when the sun, Mercury, and Venus are in a straight line, Venus will be eclipsed by Mercury, thus causing a variation in its brightness (*sol, mercurius, et venus sint in una linea recta, et per consequens magna pars veneris eclipsare per umbram mercurii*).[38] This latter arrangement is not without precedent. Most of the early Greek philosophers up to the time of Plato and Aristotle located the sun between the moon and Mercury.[39] However, Henry seems to favor placing the sun in the midst of the planets, thereby reflecting quite nicely its mean position in creation–on the fourth day, or half-way through the first week (*est medius planetarum sicut quarta feria est media septem dierum*).[40]

Disagreement over the order of the spheres might pose a serious threat to astrology, which is precisely why Henry frequently relegates ordering to the realm of the probable.[41] However, this position does not seriously undermine the prevailing astronomical systems of the day. Since the path of each planet can be dealt with separately, as will be discussed shortly, no serious problems arise if the sun is moved up or down an orb or two. There was, however, another challenge to the standard world view that proved potentially to be of greater consequence: doubting that the earth is at rest in the center of the universe.

It was commonly known in the Middle Ages that some among the ancients had developed astronomical systems that utilized a moving earth, either on its axis (diurnal rotation) or around the sun (geokineticity), but the details of these systems were, for the most part, not available. Several fourteenth-century commentators raised doubts about the earth's immobility—ostensibly for the purpose of investigating the feasibility of diurnal rotation—but little seems to have come of these discussions. Attempts to link Copernicus to medieval ancestors have not, to my knowledge, met with any success.[42] Thus, when Henry adds to his doubts about the order of the orbs a brief digression on diurnal rotation, he is raising what was a very common issue for his day, and for very medieval reasons. The problem of diurnal rotation was one of long standing duration whose discussion provided numerous opportunities to examine the prevailing theories and uncover potential weaknesses in them.

The normal occasion for discussions of diurnal rotation in the fourteenth century was the context of Aristotle's treatment of the earth's immobility in *De caelo.*[43] In contrast, Henry comes to this problem from another source, Augustine's *De Genesi ad litteram*. In this work Augustine devoted a brief chapter to the problem of "whether the heavens rest or move."[44] Within this context, the appeal for a moving earth was strong; Augustine was dealing with the "firmament," that which gives support, is firm and unyielding. Why not then give to the heavens the property of immobility and relegate mobility to the earth? Henry's rejection of this suggestion is quick and to the point. Moving the earth on its axis might account for the twenty-four-hour rotation of the heavenly bodies; however, this is only one of several motions that they have. The planets change position relative to one another; even the sphere of the fixed stars, if it is observed over long periods of time using instruments (*speciali ingenio per longas temporum revolutiones et observationes cum instrumento armillarum*), can be seen to have several motions.[45] Thus, it is obvious that merely spinning the earth on its axis does little to account for the complexity of the heavenly motions. Unlike Nicole Oresme, Henry is not willing to grant even a calculational equivalence to the two systems.[46] Just in case his brief technical arguments by-pass his audience, he adds a long list of quotations from Scripture to support his conclusion.[47]

Ultimately, from these and similar discussions regarding various aspects of the ordering and number of the celestial inhabitants, the standard medieval conception of the world emerges relatively unaltered, although with numerous restrictions as to the firmness with which some of its tenets can be maintained. Henry seems convinced that the earth does reside in the center of the universe and that the heavens are best imagined as concentric, physical orbs that stretch between the elements and the empereum. But certainty breaks down when it comes to determining in what order the heavens were set out. His caution is prompted by two concerns. First, he is fully aware of the obscurity of the heavens and is thus reluctant to set out explanations that may be known only to God. Once again he follows the inclinations of Augustine who was also reluctant to say too much about the science of the stars.[48] Second, he is moved by a deep concern for making sure that what he says conforms to the way things were ordained by the Creator. Assuming descriptions for convenience is not enough to prompt him to support a particular description. Knowledge about the world must conform to the way things are in nature. This latter attitude will have important consequences as we turn from this simple descriptive information about the heavens to a more exacting discussion of the motions of the celestial orbs.

3. Mathematical Astronomy

By far the most important task that astronomers undertook in their investigations and descriptions of the heavens was the task of accounting for and predicting the motions of the heavenly bodies. Careful observations of the heavens by the ancients had made manifest the fact that the heavenly bodies do not move at uniform rates of speed. When God distinguished the celestial orbs and set them in motion, he apparently assigned unique motive qualities to each. The lack of uniform motion in the heavens would have caused astronomers few problems had they been willing to admit that each of the heavenly bodies as an individual, as a sort of rational being, simply directs its own course, slowing down or speeding up whenever it wants to or even, upon occasion, moving in another direction. However, it seems few astronomers prior to the time of Kepler were willing to do this.[49] Perfection cannot be based on imperfection. The perfect nature and incorruptibility of the heavens cannot be based upon erratic and seemingly irrational motive powers. A more perfect order had to be found to account for the perfection of the heavens.

The most commonly accepted solution to this last dilemma was the one set out by Plato in the *Timaeus*. Plato assumed that the essence of a perfect universe is sphericity. The Creator, in bringing forth the universe,

> turned its shape rounded and spherical, equidistant every way from center to extremity—a figure the most perfect and uniform of all . . . . [And] he caused it to turn about uniformly in the same place and within its own limits and made it revolve round and round.[50]

Given this fundamental disposition, Plato apparently went on to present the following challenge to astronomers: if the essence of the universe is perfect sphericity, then the motions of the heavens must be reducible to uniform circular motion.[51] Plato's solution—or at least his suggested solution, since he did not himself go on to work out the details—was to bring perfection to the motions of the heavens by resolving them into uniform circular motions. Working out the details is what motivated Western astronomy for the next two thousand years.

The problem with this solution is that it simply does not conform to the way things are in nature. Since the planets move in elliptical orbits, their rates of speed vary as they travel through the heavens. Moreover, since many of the motions that are observed in the heavens are the manifestations of a moving earth, assuming that the earth is stationary brings in anomalies that need to be explained. The daily rotation of the earth on its axis projects out into the heavens as the twenty-four-hour, east-to-west rotation of the sphere of fixed stars. The movement of the earth around the sun makes it appear as though our planetary neighbors pursue slow, erratic, west-to-east paths that at times double back on themselves (retrograde motion) before proceeding eastward again. The fixed angle at which the earth's axis tilts as it moves around the sun makes it appear as though the sun moves higher and lower in the heavens in conjunction with the changes in seasons. And finally, the fact that the orientation of the earth's axis slowly rotates, makes it appear as though the sphere of the fixed stars undergoes a slow, and possibly oscillating, motion. These apparent anomalies, of course, can be dispensed with by moving the earth. However, since the medieval astronomer was not willing to tamper with the immobility of the earth, the anomalies took on a kind of reality. Therefore, they had to be brought into line with the rules of astronomy; they had to be accounted for with uniform circular motion.[52]

Most commonly in Henry's day this task was accomplished by employing Ptolemaic astronomy, as set out in the *Almagest* or summarized in a number of Arabic and medieval sources. Following an earlier tradition, Ptolemy (2nd. Cent. a.d.) proposed that the apparent motions of the heavens could be saved by using uniformly moving circles (eccentric circles) that are slightly displaced from the center of the universe (fig. 2) and/or a combination of circles that places the center point of one (the epicycle) on the circumference of another (fig. 3). These constructions in their simplest forms can be used to account for the changes in speed that result from the elliptical orbits of the planets. By combining the two constructions and increasing the number of circles used, a system of astronomy, Ptolemaic astronomy, can be developed that fairly well accounts for the movements of the heavenly bodies.[53]

Henry apparently had a better than average working knowledge of Ptole-

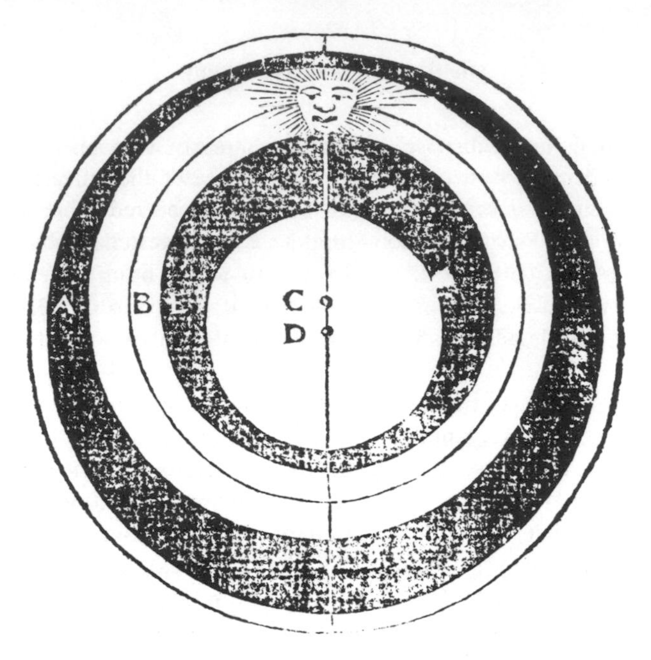

Figure 2. The Eccentric. The sun moves on its circle, B, which centers on point C, and is eccentric to D, the center of the universe.

Figure 3. The Epicycle. The eight small circles illustrate how a planet (represented as black dots) rotating on an epicycle (the eight small circles) changes position as the epicycle is carried around on the large circle upon which its center rests.

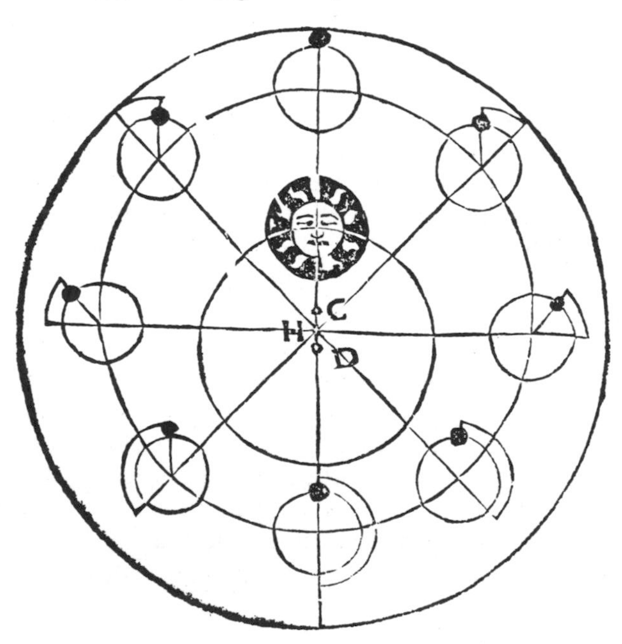

maic astronomy (mathematical astronomy was not important in the arts curriculum and most students probably never attended lectures on it) and he frequently refers to it in the *Lecturae*.[54] At the same time, he never seems to have been overly concerned with studying it in great detail. His most technical astronomical treatise, *De reprobatione*, illustrates a relatively incomplete understanding of the *Almagest*, as one scholar has recently shown.[55] The summaries that he provides in the *Lecturae* are at best inadequate, if not inaccurate. For example, at one point he notes that

> it seemed to Ptolemy that all the diverse appearances in the motions of the planets could be saved by imagining that the complete orb of any one is divided into three orbs. . . . And according to this system, if concrete being conforms to [Ptolemy's] image of the universe, the orbs of the seven planets are 29 or 30 in number.[56]

This gross oversimplification is later clarified. Henry arrives at the figure 29 or 30 by assuming that Ptolemaic astronomy uses five orbs for the moon, three for the sun, six for Mercury, and four each for Venus, Mars, Jupiter, and Saturn–an assumption that is hard to reconcile with other statements in the *Lecturae* and with Ptolemaic astronomy as taught in his day.[57] The reason for his half-hearted attempt to portray Ptolemaic astronomy accurately is fairly obvious. He does not accept it as a way to save the celestial phenomena.

Henry has two reasons for rejecting Ptolemaic astronomy. First, as he makes clear in *De reprobatione* and hints at from time to time in the *Lecturae*, there are numerous minor flaws in it that call the whole system into doubt. Most of his criticisms raised in this regard are not justifiable; however, this would have done little to diminish the effectiveness of his attack among listeners who knew even less than he did.[58] His second reason for rejecting Ptolemaic astronomy, and by far the more fundamental one, is that he basically does not believe that eccentrics and epicycles correspond to the way things are in nature (*sit dubium an . . . habeat correspondentiam in re*).[59] As he put the issue quite tersely in *Contra astrologos*, "epicycles correspond to nothing in nature" (*epicycli in re nichil sunt*).[60] This second objection proved to be a constant source of difficulty for the supporters of Ptolemy. If one is content with mathematics, fine, then Ptolemaic astronomy works well. But if a real system is sought, one that both accounts for the phenomena and corresponds to reality, then the numerous circles on circles that Ptolemaic astronomy requires stretches the imagination beyond the point of credibility. Henry was not alone in raising doubts about the correspondence of eccentrics and epicycles to reality.[61] The problem was that no compelling alternative system had been devised.

The most widely accepted alternative to Ptolemaic astronomy in the Middle Ages, one that ran a poor second as far as popularity is concerned, is

Eudoxus's system of homocentric spheres. Eudoxus (d. ca. 335 b.c.), who apparently was a student of Plato and the first to attempt a complete mathematical description of the universe, based his astronomy on a series of concentric spheres, each having its axis embedded in the one that surrounds it. By using the proper number of spheres—forty-seven or fifty-five according to Aristotle—inclining these spheres at appropriate angles, and giving them specific rates of spin, he was able to account fairly well for the celestial motions, although not as well as the results achieved with the Ptolemaic system.[62] The appeal of homocentric spheres is their readily imaginable, physical nature. The jump from the basic eight sphere to a forty-seven or fifty-five sphere universe is significant but not beyond the bounds of physical reality. However, despite its credible physical basis, Henry seems to have rejected homocentric astronomy for a third and fairly obscure astronomical system that combines features of both Ptolemaic and homocentric astronomy. This system is set out in *De reprobatione* and referred to from time to time in the *Lecturae*.

The authority from whom Henry drew the descriptive portions of *De reprobatione* is not immediately apparent. In fact, it is only by piecing together two brief references in the *Lecturae* with the earlier discussions in *De reprobatione* that the source of his astronomy emerges. The most informative statement that Henry makes about the origin of his astronomy comes near the end of the *Lecturae*, where he once again raises the problem of eccentrics and epicycles and whether they are real or not. He resolves this problem as follows:

> If someone asks me this, I send them to a certain treatise against eccentrics and epicycles, beginning "Since knowledge of inferior things, etc.," which I wrote at Paris more than 28 years ago, showing with many reasons and considerations that an astronomy of this sort [and] solely to this end is imaginary and has no correspondence to nature. I even showed there toward the end of this work, in general, by what means and how using a motion of the orbs, the appearances of the motions of the planets can really be saved.[63]

The system set out at the end of *De reprobatione*, the one that *really* saves the appearances, is a curious hybird of homocentric astronomy and an Arabic innovation introduced into Ptolemaic astronomy by Thābit ibn Qurra (fl. ninth century). Thābit attempted to account for an anomaly in the motion of the sphere of the fixed stars by attaching to it two small circles, which, when rotated, cause the eighth sphere to oscillate back and forth (see fig. 4). Henry incorporates this device into his account of the anomalistic motion of the moon.[64] His source for this added device is not Thābit, however, but Geber (Jabir ibn Aflaḥ, fl. early twelfth century) as is made clear in a second reference in the *Lecturae* to the oscillating sphere device.

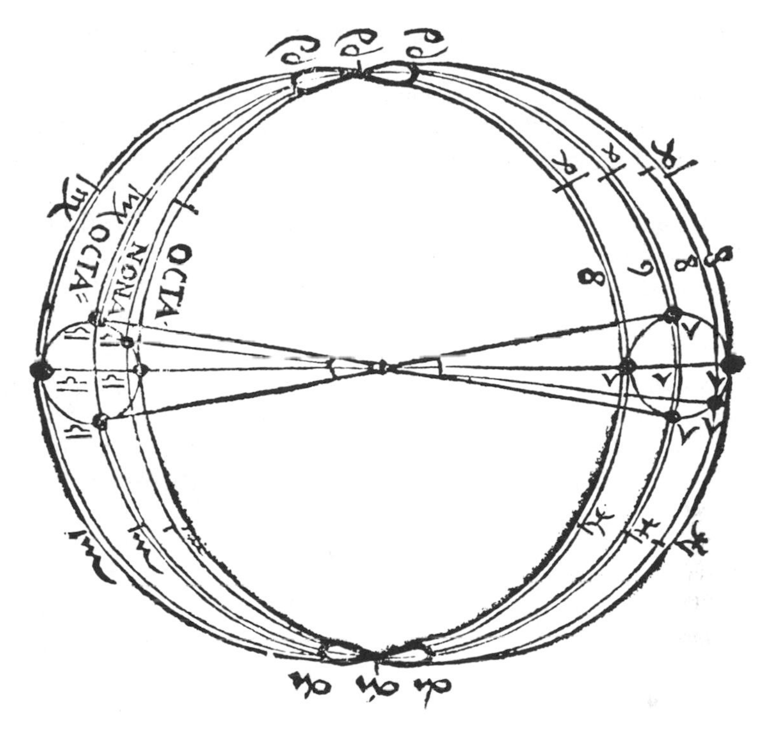

Figure 4. "Geber's" Astronomy. The two small circles at the right- and left-hand sides of the illustration rotate in opposite directions. As they do they move the sphere of the fixed stars (labelled "8" and "octa") in a looping motion (represented as a hippopede at the north and south poles) that accounts for its assumed slow and oscillating motion. Henry used this devise in his description of the motion of the moon in *De reprobatione*.

Immediately after his discussions of Ptolemaic and homocentric astronomy in the *Lecturae*, Henry describes another astronomical system

> that endeavors to save all the diversities that are manifest in the motions and dispositions or appearances of the planets with the fewest orbs. And that [astronomy] uses a particular type of extrinsic circular motion by which the sphere of the fixed stars appears to be moved in certain ways as follows: no part of the sphere of the fixed stars rests except for the center, and two points set opposite one another describe two small perfect circles, which move the epicycles latitudinally in the same way, according to Geber. Concerning this astronomy, somewhat more was said at another time in a treatise on epicycles and eccentrics.[65]

If the device described in this passage is the oscillating sphere device, which it certainly seems to be, then there can be little doubt that the source of

Henry's early astronomical ideas, as set out in *De reprobatione* and later alluded to in the *Lecturae*, is the astronomy of Geber.[66] Henry's search for reality in mathematical astronomy has thus driven him to reject the two standard astronomies of his day and accept in their place an astronomical system that had apparently very little circulation throughout the Middle Ages.

In his pursuit of this particular approach to astronomy, there can be little doubt about Henry's motivation. He is clearly not interested in accurate mathematical calculations. He did not strive to attain such in *De reprobatione* and, in fact, suggests in the *Lecturae* that their attainment may not be possible.[67] Instead, the entire thrust of his astronomy, both mathematical and physical, is directed toward describing and accounting for the way the heavens are really (*in re*) disposed. The orbs are, for Henry, real entities that must be taken into account. Just as the physical astronomer needs to consider how elemental forms can be moved through the orbs without fracturing them, so too, the mathematical astronomer must be certain that his constructions can exist. They cannot be figments of a mathematical imagination. Why he should have had this overbearing concern for reality will become more apparent when we return to a consideration of the influence the heavens have on the elements (chap. VI), after discussing (chap. V) how the elements are disposed.

CHAPTER V

Day Three: The Elements, Cosmography, and Meteorology

> *And God said, "let the waters that are beneath the heaven be drawn together in one place, and let dry land appear." And it was made in this way. And God called the dry land, earth. And the assembled waters he called seas. And God saw that it was good. And he said, "let the earth bring forth upon the earth vegetation that is green and bearing seed, and the fruit tree bearing fruit according to its kind and the seed that is in it." And it was made in this way. And the earth produced vegetation that is green and bearing seed according to its kind, and the tree bearing fruit, and each having seeds according to its kind. And God saw that it was good. And the evening and morning were made, day three. Genesis 1:9–13.*

Once the celestial orbs were distinguished and provision made for the stars and planets that will appear on the fourth day, God turned his attention to the elements that lie below the stars in the sublunar or lower (*inferior*) region. The organization of this region had been established *in principio*. Even before they were formed, the elements were potentially arranged in four concentric spheres, extending out from earth through water and air to fire. On the third day this potential organization was made actual, as the elements were formed and given precise dispositions in the sublunar realm. At the same time that he formed the elements, God took up the role of farmer, sowing the seeds of plants in the earth and initiating the growing process. But plants were not completely formed on the third day. Their growth depends in part on celestial influences whose source has yet to be created. Accordingly, the discussion of plants is best left to a later chapter where other living creatures will be discussed (chap. VII), thereby allowing the present chapter to pertain exclusively to the elements.

The consideration of the elements comes under the domain of a number of sciences. Their matter-form composition has already been discussed in conjunction with metaphysics (chap. II). Their relevance to physical astronomy was mentioned as part of the work of the second day (chap. IV). From here one could go on to consider how their qualitative dispositions account for the various interactions they undergo in producing mixtures and complex bodies, as taken up by the study known as "the science of generation and corruption." But before the qualitative interactions of the elements are mentioned, two sciences that pertain more directly to the work of the third day need to be discussed–cosmography and meteorology. Both of these sciences consider the elements as elements, focusing particularly on how they are disposed and move in the sublunar region but not on what happens when they react with one another to produce compounds and mixtures. The latter aspect of the "elemental" sciences will be taken up in the next chapter in conjunction with the work of the fourth day.

1. Sublunar Cosmography

Cosmography, the science that considers the physical organization of the universe, comprised an important part of medieval science.[1] Its relevance to all types of studies, including religious studies, had been noted as early as the time of Cassiodorus (d. ca. 575), who recommends to monks that "you ought to acquire some notion of cosmography, in order that you may clearly know in what part of the world the individual places about which you read in sacred books are located."[2] The tradition available in the Middle Ages to pursue this advice was long and varied. It extended from the ancient works recommended by Cassiodorus, such as Julius Solinus's (*Polyhistor* fl. 250 a.d.) elementary *Collectanea rerum memorabilium*, through even more popular summaries of the writings of the ancients contained in Pliny's (d. 79 a.d.) *Naturalis historiae*, Isidore of Seville's (d. 636) *Etymologiae*, and other encyclopedias; and through the occasional cosmographic treatises written during the Middle Ages, such as Honorius of Autun's (?) (fl. early twelfth century) *De imagine mundi* or the *Liber floridus* of Lambert of St. Omer (d. ca. 1125), to works written by Henry's own contempories at the University of Paris.[3]

Despite its importance, cosmography *per se* did not play an important role in the science associated with medieval universities. Perhaps this is largely due to the fact that Aristotle's interest in the disposition of the elements tended to focus heavily on theory rather than on descriptive details. His extended treatment of the sublunar region in *De caelo*, books three and four, is essentially a discussion of the properties of the elements, not their geographic distribution. Somewhat more detail on specific phenomena is set out in *Meteorologica*, although again without much concern for specific examples and places. Accordingly, commentators on these and related works, such as Sacrobosco's *Sphaera*, had no obvious reason for bringing the wealth of detail found in the more popular cosmographic tradition into their highly technical discussions of the elements. When they did, excuses for going beyond Aristotle's interests had to be manufactured.[4] There is, therefore, a marked difference between cosmography as set out in the popular and encyclopedic tradition of Pliny, Isidore, and others, and in scholastic treatises on the sublunar region. Henry, for example, displays very little concern for the location of places, land masses, oceans, and so forth. He is more concerned with theoretical speculations regarding how God went about the process of distinguishing than with a description of the end product of the third day's work. Indeed, it is only by piecing together many incidental statements made throughout the *Lecturae*, as will be done later in this section, that any understanding of his conceptualization of the *mappamundi* emerges.

The importance of the theoretical or speculative side of cosmography from a scholastic standpoint is obvious. The way things are organized in the sublunar region depends on the characteristics of the things that are being organized. The drawing together of the seas and dry lands, for example, is a process that is governed by the properties of water and earth. What is of initial importance, then, and what comprises the bulk of speculation in scholastic treatises relating to cosmography, is how one gets from the general properties of the building blocks of the universe to their final disposition. This means that the starting point for a discussion of the sublunar region must be a description of the four elements, earth, water, air, and fire, and their properties.[5]

A. Elemental Properties, Gravity, and Levity

Tradition and authority firmly established for the medieval scientist the properties that characterize each of the elements.[6] The specific moisture-temperature and motive qualities assigned to each by the ancients were accepted as obvious givens that perhaps needed some clarification but that were not to be replaced. Thus Henry assumes, and on several occasions carefully explains, that earth is dry and cold; water is cold and wet; air is wet and hot, and fire is hot and dry.[7] In addition, each element has a natural tendency to move toward its proper place in the universe; earth, placed in the sphere of water or air, moves down (*deorsum*); fire, placed in the sphere of air or earth, moves up (*sursum*), and so forth.[8] In combination, these two sets of properties, the moisture-temperature qualities and the motive qualities, are what characterize each of the elements as a particular element.

Of the two sets of properties, the one that has the most immediate bearing on cosmography is the one that encompasses the motive qualities. These qualities explain why the sublunar region is structured in four concentric spheres. Earth is at the center of the universe because that is its proper place. If it were placed anywhere else in the universe, it would move locally and rectilinearly down (*deorsum*) until it reached its proper place. Why? Because that is its proper nature; that is the way it was created. Just as to be a pear tree means to have the capacity to bear pears and to be an apple tree means to have the capacity to bear apples, to be earth means to reside naturally in the center of the universe and to seek out that residence if displaced from it. Part of the creative act of informing the world is to give to the elements motive qualities that account for their natural movements within the sublunar realm.[9]

Gravity and its opposite, levity, thus become, in the medieval world, qualities that reside within bodies. A rock falls to earth not because something external to it acts upon it and draws it toward earth, but because that which is natural to it, its own proper nature, inclines it toward its assigned place in the universe. From the standpoint of modern science, this of course is not really a satisfying explanation for gravity. For the medieval scientist, however, it was. Form brings to objects many characteristics. It makes matter into something that is extended, a body;[10] it determines that fire is hot, water wet, and so on. Why then cannot form also make bodies move locally? Besides, if this explanation of gravity is accepted, a very ticklish problem is avoided, the problem of action at a distance.[11] If gravity is not attributed to an object's proper nature, then some other mechanism must be found to explain how one body somehow mysteriously affects and acts upon a distant body. For a medieval scientist, any explanation of this sort would have been less satisfactory than assuming an internal mover for two reasons: First, since the medieval scientist had no way of determining how distant bodies affect one another it is still only explanation by definition; and second, it violates a number of basic principles, not the least of which is that mover and moved must be in direct contact.[12]

Internalizing the *ratio* behind falling bodies and making it a proper nature did not deter the medieval scientist from attempting to describe gravity in more detail. For example, experience clearly indicates that the velocity of a falling body increases over the course of its fall. In what proportion was still uncertain. Both time and distance were taken into account as factors, and there was no certainty as to which, if either, is the more directly related.[13] But why the rate of descent increases during a fall is another matter. A number of explanations were given to account for this, including one that brings into focus a very important fourteenth-century theory, namely, the impetus theory.

The impetus theory, as initially derived, seems to have been aimed at explaining why an effect continues to be operative even after its cause is no longer in direct contact with it.[14] Why, for example, does a rock (a projectile) continue to move through the air once it leaves the hand of the thrower? Aristotle's answer to this question placed the burden for continued motion with the air (what else is there that remains in contact during flight?). He reasoned that either the air that is pushed away from the rock by its forward motion swirls around and, in turn, pushes the rock from behind; or, that as the rock is thrown, the air surrounding it is moved, and in turn this air moves the next piece of air, and so on, and in the process carries the rock along with it.[15] For several obvious reasons, not the least of which is the problem of assuming that air moves rocks, both of these explanations were rejected by a number of Aristotle's followers, including some of Henry's predecessors at the University of Paris, and replaced by the impetus theory.[16]

In line with the general tendency of medieval natural philosophers to shy away from external mechanical explanations, the impetus theory displaces the source of projectile motion from the realm of external cause and effect and internalizes it within the projectile itself. (In this sense, it is even more Aristotelian than Aristotle's science.) That is to say, instead of assuming that a projectile moves because of some external mechanical action, supporters of the impetus theory postulated that the process of throwing imparts to an object an additional qualitative disposition (an additional form or nature), which in turn accounts for the object's continued motion. The rock continues to move along its trajectory because it has, once it leaves the hand of the thrower, "a quality naturally present and predisposed for moving the body."[17] Its nature now becomes that of an object moving along a path that had previously not been natural to it. As it moves the projectile, this added quality is weakened, either because it is not a permanent quality and therefore naturally remitted, or because of the resistance of the air. As a consequence, the projectile slowly loses its imparted motion and assumes its natural motion, that is, toward the center of the universe. Consequently, a rock thrown away from the earth falls through an arched trajectory back to the earth, as experience easily confirms.[18]

The impetus theory proved so successful to the minds of fourteenth-century scientists that it was applied to a number of phenomena besides projectile motion, from the problem of celestial movers (see chap. VI, sec. 1) to that of the Eucharist.[19] What is important for the purpose of this chapter, is that John Buridan, a leading spokesman for the impetus theory at Paris, applied it to the problem of falling bodies. Buridan reasoned that natural gravity, what Henry calls the natural tendency of a body to move down due to the proper nature of earth, cannot of itself cause a body to move more swiftly as it falls. Therefore, some added motive quality (impetus) is needed to account for its acceleration.

> This impetus has the power of moving the heavy body in conjunction with the permanent natural gravity. And because that impetus is acquired in common with the motion, hence the swifter the motion is, the greater and stronger the impetus is.[20]

Downward motion thus becomes a product of two motive qualities, according to Buridan: one, the natural inclination of heavy bodies to move down, and the other, the added increments of motive quality, called impetus.

While not directly confronting the problem of the acceleration of falling bodies, Henry's teachings on proper and common natures seem to eliminate any need for Buridan's explanation of acceleration. Henry regards gravity as something that is entirely natural to an element through the action of its form.

> There is no need to have certain special motive qualities placed in the elements, such as gravity and levity, because . . . the form of earth . . . can in a simple manner cause downward motion.[21]

Gravity is not, therefore, a special proper nature that is added to bodies. Neither is it a common motive quality in all elements, moving some up and some down. If gravity were a common motive quality, it would be no more natural to earth "than the natural motion that depends for its cause on the quality that they call impetus."[22] Impetus is thus, for Henry, an unnatural quality that is distinct from natural qualities such as gravity.

Arguing that gravity is a natural and impetus an unnatural motive quality does not, of course, exclude impetus from an explanation of falling bodies. Buridan used both qualities; one to explain downward motion, the other to explain acceleration. However, the whole point of Henry's critique of causality, as it appears in his discussion of the role of proper and common natures in explanations of gravity and in his more general simplification of the entire chain of events in the sublunar region, is to eliminate the need for added causes when the simple forms of elements and their primary qualities will suffice. Therefore, even though he uses the impetus theory in other contexts, it seems that he did not employ it in the case of gravity, and its opposite, levity. The elements, through the action of their forms, tend to seek their respective spheres in the universe; no further explanation of how they fall or rise is needed.[23]

The end product of the actions of the forms of the elements should, of course, be a perfectly layered, perfectly spherical, sublunar region with each element in its proper place. This is not, however, the way the world is ultimately disposed. Therefore, additional factors need to be brought in to explain how God gathered together the waters and made the dry lands emerge. Such are the proper considerations of cosmographers, both scholastic and popular.

B. Disposing the Elements

Many opinions circulated in Henry's day regarding the manner in which the waters and dry lands were separated. One school of thought on this subject suggested that earth (the element, not the planet) somehow protrudes out through the sphere of water, thereby accounting for the mountains, islands, continents, and other land formations that we observe. But how does earth, which naturally occupies a position within the sphere of water, protrude out through that sphere? Henry mentions two of many explanations that were commonly given to account for this: (1) The earth could be slightly off center, thereby causing its sphere to extend out through the sphere of water. This, he suggests, would be possible if some permanent celestial power (*per quandam rigitivam caelestem virtutem*) held the earth off center.[24] (2) The earth could be shaped more like a triangle than a perfect sphere, again allowing a portion of it to extend out through the waters. However, this would once more require a portion of the earth to remain outside its normal sphere. In this case, Henry accounts for the anomaly by noting that if the earth were once made triangular, it would stay that way because it is solid and not fluid, and that if the displaced earth were porous, it would absorb air and heat, thus making it lighter than normal earth and counteracting its normal tendency to return to its natural place.[25]

Another school of thought confronting the same problem placed more credence in the Biblical account of the drawing together of the waters and sought to explain how God either drew back or otherwise removed the waters from the earth. Again, several ways were advanced to account for this: (1) Some of those choosing this approach followed Plato's suggestion that the earth is filled with pores and cavities and assumed that these pores and cavities draw in the waters and thus remove them from portions of the earth.[26] (2) Others followed Aristotle's lead and assumed that the sun dries up portions of the waters and leaves dry land exposed. This simple explanation would be easily acceptable were it not for the fact that the sun has yet to be created.[27] (3) Still others believed that through some influence, either celestial or divine, the waters are drawn back and held behind a specific boundary, thereby causing them to pile up in some places and be absent from others.[28] This pile of waters, it should be noted, supplies a ready source for the waters of the Deluge.[29]

Deciding which, if either, of these two approaches to the disposition of the waters is to be accepted was no simple task. Evidence strongly suggests that the dry lands do project up above the sphere of water. How else is one to explain the fact that major rivers, such as "the Rhine, Danube, and Rhône" (*Renuss, Danubius, et Radanus*), originate in the mountains and flow to the seas?[30] However, there is also evidence to suggest that some lands lie below the oceans–for example, the lands between the Red and Mediterranean Seas, which were threatened with flooding when an attempt was made to connect the two with a canal, or the more obvious example, for Henry, of the Low Countries (*Zelandia*).[31] Not surprisingly, then, his own cautious solution to the problem of disposition is a composite of several opinions.

According to Henry, the earth began as a spherical, nonporous body that was completely surrounded by water. On the third day God turned this perfect sphere into a porous, mountainous, sponge-like body (*facere ipsum montosum, porosum et cavernosum ad modum cuiusdam magnae spongeae*) that absorbs a portion of the waters, thereby allowing the dry lands to be exposed. This explanation is not, however, of itself complete. Had this been all that God did, it would seem that the dry lands ought to appear at random over the surface of the waters, even at a point opposite where man lives (the so-called antipodes)–a suggestion that Henry clearly rejects. He therefore argues that in addition to absorbing the waters into the caverns of the earth, the land is displaced from the center, thereby creating deep oceans in one hemisphere and a habitable region in the other. The amount of eccentricity of the earth is quite small (*pars terrae arida non elevatur notabiliter multum per modum montis*), as is evidenced by the appearance of its shadow during eclipses. Still, it is sufficient to account for the drawing together of the waters and the appearance of the dry lands.[32]

The consequences of this conception of the world are several and significant. First of all, the fact that the sphere of water completely surrounds the sphere of earth leads to the assumption that there is considerably more water in the universe than earth, and so on for air and fire. Concerning this matter, Henry gives two opinions: one, supposedly originating with Aristotle, assumes that there is ten times (*decuplo*) as much water as earth, ten times as much air as water, and ten times as much fire as air;[33] the other begins with data given by Ptolemy and derives a ratio for the amount of each element of 1 to 33 to [1,100] to 39,000.[34] Furthermore, since the displacement of the earth from the center is slight, the amount of land extending out through the waters is small in area. Again, Henry makes an attempt to calculate this. Using the basic fact that the dry lands have at their greatest extension a line extending in latitude from the meridian to the North Pole (that is, one quarter of the circumference of the globe), he reaches the conclusion that they cover scarcely one-seventh of the surface of the earth-water sphere.[35] Finally, the fact that the waters almost completely surround the dry lands suggests that

> a ship is able to set sail from the western part of the dry land, go over the water in the lower hemisphere, and cross under our feet in returning to the eastern part of the dry land,[36]

a suggestion supposedly demonstrated to be true a century later by Columbus.

The eccentricity of the sphere of earth in relation to the sphere of water is the only major anomaly found in the sublunar region. The sphere of air, which comes next in the order of the elements, completely surrounds the sphere of water and is in turn completely surrounded by the sphere of fire.

The mean position of air between the extremes of the fiery region and the two lower elements commonly led to the assumption that its sphere is further divisible into three subregions: a lower region that has the proper qualities of hotness and wetness, a middle region that is cold, and an upper region, called the "*aestus*" by Albert the Great (d. 1280), that is overcome by the influence of fire and therefore hot.[37] The importance of these three subregions will become more apparent in the following section where the meteorological phenomena associated with the elements are discussed.

The final elemental sphere, the sphere of fire, is seldom mentioned by Henry, possibly for good reason. Its invisibility frequently led commentators to question whether such a sphere actually exists. Fire, after all, has the obvious property of giving off light; therefore, if a sphere of fire exists, we ought to see this as a sphere of light in the heavens. A predictable reply to this argument, to slip for a brief moment into a common medieval manner of argumentation, would be to deny the antecedent to the conclusion, that is, to deny that fire necessarily gives off light. To be fire simply means "to possess the capacity to give off light," not "to give off light." Light is not a proper nature of fire, as are levity, hotness, and dryness. Consequently, fire has no more necessity to give off light than someone who has the capacity to laugh must laugh.[38] Besides, as Henry had himself pointed out, the flames that we perceive as fire are not pure fire but a mixture of fire and light.[39] So there are a number of reasons that could be given to account for the fact that there is a sphere of fire, which was seldom if ever doubted, even though we do not see it.

Having laid out the general disposition of the elements, a complete treatment of cosmography would go on to a discussion of "rivers and seas and regions and some (perhaps only a few) cities."[40] It is at this point in his pursuit of cosmographic concerns, however, that Henry's attention to detail begins to flag. He makes brief reference to the five climatic zones during a discussion of the sun's path through the heavens, thus suggesting that he follows the standard teaching on this subject. It was generally thought that the earth's surface (the surface of the earth-water sphere) could be divided into five zones: a central, torrid zone; two cold, polar zones; and two intervening, temperate zones.[41] But as for "rivers, seas, and cities," he has very little to add—very little, that is, with the exception of one obvious section of Genesis that called for a discussion of places, the section that mentions the location of Paradise (Genesis 2:10–14).

The account of Paradise in Genesis centers on the four major rivers of the world (the Ganges, Nile, Tigris, and Euphrates Rivers), which flow from the "River of Eden." In his discussion of these rivers and the lands through which they flow, it is clear that Henry relies heavily on Isidore's *Etymologiae*. Therefore, his conception of the geography of the world is basically the same as that set out by the ancients and more or less followed throughout the Middle Ages. He assumes the standard threefold division of the habitable lands: Europe, Africa, and Asia (fig. 5).[42] Asia was generally considered to be the largest of the land masses, occupying fully one-half of the habitable surface, while Africa and Europe occupy the other one-half according to Isidore. Consequently, the Ganges, Tigris and Euphrates Rivers all refer to rivers of Asia, the first being in India, while the second and third border the land known as Mesopotamia. The fourth river, the Nile, flows through Africa,

Figure 5. *Mappamundi.* In addition to the three continents (Europe, Africa, and Asia), this early fifteenth-century map locates major cities, rivers, and countries (e.g. Rome and Paris; the Rhine, Danube, and Ganges; and Germany, Italy, and Egypt). The Mediterranean can be seen dividing Europe from Africa and running from west to east or from the bottom to the mid-portion of the map.

bounding on its one side the scorched and monster-ridden land of Ethiopia.[43] As to the third land mass, Europe, Henry has very little to say. He mentions only a few local place names in the *Lecturae* and these are, for the most part, contained in general references that provide no insight into the details of his conceptualization of the map of Europe.

The absence of specific details from the cosmographic discussions of medieval scientists lends credence to the commonly held assumption that their narrow and orthodox outlook allowed little time to focus attention on distant horizons. Such is certainly true as far as the written word is concerned. Even Henry's younger contemporary, Pierre d'Ailly (d. 1420), who wrote a number of works on cosmography and related topics, seems to have shown little interest in the developments and explorations that were going on around him.[44] But it must also be borne in mind that this conclusion is based on a genre of literature that demanded speculation about the generalizations and theories that lay behind the dispositions of the elements, without asking for particular instances and individual places. What we have been discussing, therefore, is as much or more the product of a method of inquiry as it is the product of the tastes of the age. That is to say, the absence from scholastic treatises of tales about the Orient and Africa, which had become so much a part of the European mentality after the Crusades and journeys of men such as Marco Polo,[45] does not mean that scholastics ignored or were ignorant of what was going on around them; it simply means that the applicability of such tales to their science had not yet become apparent.

The importance of comsographic concerns to scientific matters will, of course, become apparent to succeeding generations. Cosmography will slowly make its way into the classification of the sciences over the next hundred years, as will other practical disciplines such as mining, navigation, and the military arts.[46] Even standard scholastic texts, such as Sacrobosco's *Sphaera*, will come to be interpreted as encompassing more of the particulars relating to the sublunar region than are found in medieval commentaries on the same treatise.[47] Renaissance science slowly brings to the domain of *scientia* practical endeavors that were previously thought by most to be irrelevant to science. Herein lies one of the greatest differences between Henry's science and that of his Renaissance counterparts a hundred or so years later. What impact this broadening of the domain of *scientia* ultimately has on the advance of science happily falls beyond the scope of this book. Presently scholars are deeply divided as to the importance of the fusion of the practical with the speculative in the Scientific Revolution. What should be clear from this discussion of cosmography is that in fourteenth-century Paris and Vienna this fusion had still not taken place. The limits of a complete understanding of the physical world had yet to be reached. There was no reason, therefore, to turn to other endeavors before the first had been completed.

2. Meteorology

Meteorology, as defined by Albert and others prior to Henry, deals mostly with the movements of elements from sphere to sphere and the subsequent phenomena that result from these movements: rain, hail, comets, rivers, volcanoes, and so on. As such, Albert made an attempt to subdivide this science into two branches: one that deals specifically with the movement of the elements from sphere to sphere as covered in books one through three of Aristotle's *Meteorologica*, and one that concerns the qualitative interactions of the elements as covered in book four of *Meteorologica.*[48] Including book four of the *Meteorologica* in a discussion of meteorology led to problems, however, since the qualitative changes associated with the elements treated therein are also discussed by Aristotle in *De generatione et corruptione*. Such problems led Aquinas to suggest that the content of book four of *Meteorologica* is best taken up in conjunction with a discussion of the generation and corruption of the elements.[49] This suggestion is utilized in the present section, thereby keeping meteorology as the science that deals exclusively with the movement of the elements from place to place and not their qualitative interactions.

How phenomena result from the movement of the elements from sphere to sphere is perhaps easiest to understand in the case of water. The delicate balance that exists between rivers, rain, and the large bodies of water that cover the surface of the earth is obvious. It provides, according to Henry, evidence to support the famous contention made in Scripture that "God disposed all in the universe according to number, weight, and measure" (Wisdom 11:21).[50] Each part of the universe is balanced with every other part, even as change takes place. The rivers that flow into the oceans do not cause them to overflow. Instead, their input is balanced by the escape of vapors into the air, leaving behind, according to a widely accepted theory, an earthy residue that accounts for the salinity of the ocean's waters (*in salsedinem revertente*).[51] The desalinated waters that evaporate from the surface of the oceans rise into the cool middle region of the air where they are drawn together as clouds and then condensed, whereupon they fall to the earth as rain. The rain, in turn, becomes the source of rivers and springs, which flow into the oceans and begin the process again.[52] And so the cycle goes on, with water constantly circulating between its own proper sphere and the spheres of earth and air, in the process watering God's creation and providing for man.

The evaporation-condensation cycle that accounts for the circulation of water was applied by Aristotle to a number of meteorological phenomena besides rain, including dew, hoarfrost, snow, and hail. Dew and hoarfrost are phenomena that take place in the lower region of the air. Dew results from water vapors that are condensed in regions where there is not sufficient warmth to turn them back into vapors but sufficient warmth to keep them from freezing. Dew that freezes is hoarfrost. Snow and hail form in the cold, middle region of the air–snow being frozen vapor and hail being frozen rain. The latter distinction is not as simply maintained as might be imagined. Snow falls in winter, whereas hail is more commonly associated with warmer seasons. What causes rain to freeze when it is warm? In response to this question, Aristotle rejected the suggestion that the freezing of rain is due to the movement of a cloud to a higher, colder region and posits, instead, that it is caused by the greater coolness of the middle region of the air in warmer seasons. The middle region of the air is cooler in summer because the concentration of heat in the lower region leaves less heat in the middle region, hence rendering it cooler and capable of giving rise to hail.[53]

The simple evaporation-condensation cycle with its many variations accounts for only one portion of the movement of water in the sublunar region. The earth, it will be recalled, is not a perfectly round, impenetrable solid; rather, it is made on the third day into a porous, sponge-like body that absorbs a portion of the waters. The waters that are absorbed into the body of the earth perform a very inportant role in accounting for additional meteorological phenomena.

(a) The waters that fill the pores in the earth's crust (*subterranei meatus*) at times and in certain places bubble forth as springs, which in turn become sources of rivers. The pores assist in this process by providing either a site for the condensation of vapors or an outlet through which waters that are collected in other places, such as high in the mountains, can flow. At times the earth can even act as a sponge and draw up waters that are deep within the earth, again producing springs.[54]

(b) The pores in the earth's crust also serve a reverse function, allowing waters collected on the surface of the land to return to the oceans. This means that water undergoes a second, underground cycle that, along with the evaporation-condensation cycle, helps keep the elements in balance.[55]

(c) Some of the waters that collect in particularly deep and obscure pores (*in occultis meatibus subterraneis*) are influenced by the generative powers of the earth and changed into metals and other precious stones.[56] The shift, in this instance, from a phenomenon involving movement from place to place (seepage of water into the deep pores) to one that concerns a qualitative change illustrates how close meteorology and the science of generation and corruption are in content. One science leads to another with the boundary between them at times being hard to delineate.

The hot-wet vapors or exhalations (as they were also commonly called) of evaporation are not the only vapors that issue forth from the earth. Following Aristotle's descriptions and popular belief, medieval meteorologists conceived of the ground on which they dwelled as a kind of sweating, fuming mass that is constantly exuding all sorts of exhalations. This prompted speculations about a second type of exhalation–a hot-dry exhalation that, as it travels out through the pores of the earth and through the atmosphere, gives rise to other meteorological phenomena.[57]

(a) Comets were commonly believed to be hot-dry exhalations (*exhalationes calidae et siccae*) that travel up through the air and burn when they come in contact with the hot-dry qualities of the sphere of fire.[58] This is only one of seven explanations of comets that Henry gives, but it is clearly the one that he accepts. Comets are not, therefore: (1) wandering stars that rarely appear, (2) partially made up of heavenly matter, (3) rarified rays of the sun, (4) subtle posphorescent bodies that are raised from the elements and shine by themselves, (5) part of the air and fire that is made luminous, or (6) a miracle caused by God.[59] Comets are entirely natural phenomena that occur in the sublunar region and involve only one type of exhalation, a hot-dry exhalation that issues forth from the earth.

(b) The mechanism used to explain comets was extended by Aristotle to include shooting stars and the Milky Way (*galaxia* or *via alba*). Shooting stars arise when combustible, hot-dry exhalations are either injected in stream-like fashion into the fiery region or, once in the fiery region, moved circularly by the friction of the adjoining lunar sphere and simultaneously ignited along the length of their path.[60] The Milky Way results from numerous exhalations being drawn together in that portion of the heavens where the stellar influences responsible for their elevation are the strongest–along the band of the zodiac–and then ignited.[61] Aristotle's explanation of the Milky Way was rejected by Albert, who is followed in this matter by Henry. Albert argues at some length that the Milky Way is not a sublunar but a heavenly phenomenon resulting from the combined light of many small stars that are collected in a single orb.[62] Technically, then, the Milky Way should not be included in a meteorological discussion; however, Aristotle's organization, if not his theory, usually held sway, thus keeping the discussion of the Milky Way within the context of meteorology.[63]

(c) The hot-dry vapors that give rise to comets and shooting stars in the fiery region produce an entirely different set of phenomena when confined to the lower region of the air; here they give rise to the winds that blow across the surface of the earth.[64] By carefully bringing into consideration the cooling effect of the middle air and polar regions, the warming effect of the tropics and sun, and various special influences such as the streams of extremely cold air that shoot out from the earth (*voragines terrae de quibus spirant venti gelidi*),[65] the medieval meteorologist was able to account not only for winds in general but for all of the special prevailing winds, such as the North and South Winds, and the other winds mentioned by the ancients.[66] Some of these discussions become quite complex and draw on a great deal of common experience. Almost always, however, the observations that are mentioned are sorted out in accordance with a few standard assumptions that underlie all meteorological phenomena.

The substance of such assumptions should by now be fairly obvious. The primary driving force behind the movement of the elements is temperature variation. Heat produces vapors and causes them to rise; cold condenses vapors and causes them to descend. With the exception of those movements that are caused simply by gravity and levity, everything of a meteorological nature that takes place in the sublunar region arises from wet- or dry-hot exhalations that are driven from sphere to sphere by heat and cold. The driving force behind heat and cold lies outside the sublunar region, in the heavens. The sun is obviously the major source of heat in the universe. Its tendency to concentrate heat in certain regions, as in the lower region of the air, leaves other parts of the sublunar region cold. So heat and cold derive from the heavens, thus placing the action of the elements ultimately in the hands of the heavenly bodies. These, however, have yet to be created. And so it would seem that on the third day the elements stand poised and ready to move but as yet unmoved, since their movers will be created on the following day.[67]

3. Geological Variation

Before going on to the work of the fourth day, there remains one fascinating problem regarding the work of the third day that is worth a few brief paragraphs. What has been described to this point concerns the disposition of the elements as they appear to us at the present time. But can it be said with certainty that the way the earth appears today is the way it was created on the third day? Henry at no time seems to hold to any concept of geological evolution, so, in this sense, the past and present dispositions of the elements ought to be much the same. However, two past events stand in the way of maintaining geological continuity—the Fall of man and the Flood. The latter event has obvious significance vis-à-vis the work of the third day; it involves the total, albeit temporary, destruction of the established order of nature, since the waters trespassed across their assigned boundaries and completely covered the earth. This leads one to wonder: When they receded, did they leave the disposition of the elements exactly as it was before?

An answer to this last question is not easily given. Since it is not known, according to Henry, how the waters receded, there is no way to determine exactly what part of the lands became dry first or whether the part that did become dry was the same part that was dry prior to the Flood.[68] Thus, it is possible that the orientation of the dry lands and oceans may have been altered as a result of the Flood. However, we do know from Scripture that mountains and valleys existed prior to the Flood. It would seem, therefore, that geologically the earth was similarly disposed before the Flood, even though its geological formations may have been oriented differently.[69] But this still does not get us back to the third day.

An argument can be advanced to support the assumption that if the disposition of the elements was not significantly altered by the Flood, then the way the elements are disposed now is the way they were disposed on the third day. God's natural creative acts ended on the sixth day; therefore, there is no other time that he could naturally have made mountains and valleys. As convincing as this argument might seem, especially for someone such as Henry who leans toward natural causation whenever possible, he does not accept it. Instead he posits that

> the earth, from the beginning, was uniform throughout its parts and neither rocks nor minerals were in it. . . . God did not make mountains from the beginning but secondarily, after the Fall of man, and so too many other things. . . . [70]

It would seem, therefore, that according to Henry there have been two states in which the earth has existed, a primordial state (*status primordialis*) and a final state (*status finalis*).[71] What we have been discussing to this point has been the final state, the one that comes about mostly by the design of the Creator but also in part as a result of the sins of man.

God's interference with the primordial condition of things at the time of the Fall should not lead to the suspicion that the world is unstable and constantly subject to the will of the Creator. The Fall is an extraordinary event in the history of the world, one whose consequences far exceed any other disruption in the normal order of things. Moreover, the disposition that results from the Fall becomes ordained in nature and therefore in the natural state of the things that are primarily the object of scientific inquiry. Finally, even in acting miraculously at the time of the Fall, God did not by-pass nature; he acted through nature (*naturali cursu*).[72] How God acts through nature will become clearer when we have examined, in the following chapter, the golden bond that ties the two together.

CHAPTER VI

Day Four: The Stars, Physics, and Astrology

And God said, "let lights be made in the firmament of the heaven. And let them divide the day and night, and let them serve as signs and for times, both days and years, as they shine in the firmament of the heaven and illuminate the earth." And it was made in this way. And God made two great lights–the greater light to rule over the day, the lesser light to rule over the night–and stars. And he placed them in the firmament of the heaven to shine forth upon the earth, and to rule over the day and night, and to divide the light from darkness. And God saw that it was good. And the evening and morning were made, day four. Genesis 1:14–19.

Strictly speaking, the work of the fourth day pertains to the heavens. During the first three days the outlines of the universe were sketched out in the amorphous mass that was to become the universe. Now, on the fourth day, the process of filling in the details (*opus ornatus*) is begun. God commences perfecting the design that has to this point been rather crudely roughed out. As with the work of distinction that preceded, ornamenting the universe begins in the heavens. God sets out two great luminaries, the sun and moon, to rule respectively over day and night. These are accompanied by the other heavenly lights that take up residence in their respective orbs and shine forth upon the earth. Together, they complete the celestial design. Except for occasional and miraculous additions,[1] the work of the fourth day sees the heavens established in the form in which we know them.

Notwithstanding its narrow limits, the work of the fourth day is used by Henry to delve into topics that extend well beyond celestial phenomena. He had already raised a number of more technical issues regarding the heavens in conjunction with the work of the second day, as drawn together and augmented in the discussion of physical and mathematical astronomy (chap. IV). Now, what interests him is not stars *qua* stars so much as stars as they influence the elements and direct the creatures that lie below. This interest engages him in what is essentially a twofold project: first, elucidating the manner in which celestial causes lead to earthly effects, and second, determining whether men can predict these earthly effects by observing the objects

Of the four types of cause that properly fall under the domain of physics (efficient, formal, material, and final), the one that initially has the most bearing on the work of the fourth day is the efficient cause. The stars perform a crucial role in helping the universe run smoothly and according to God's plan. They comprise one of the most essential and obvious links in the great "golden chain" that joins heaven to earth and men to God. None of God's agents (the intelligences), Henry argues, can exert their influences "in the inferior regions unless [they act] through the mediating influential qualities of the celestial orbs and stars."[15] This means that every effect that is observed in the sublunar world is at least in part and in some way initiated by (has its efficient cause in) the stars. What we now must do—and this is the major task that Henry set out for himself in his well-known treatises on cause and effect—is trace the path of the chain of efficient causes that move the universe from God, through the stars, and into the elements.

At the beginning of the golden chain comes God. He is the first link and activates all that follows. He is both Creator and First Mover. First he created all that is movable and subject to change (*omnia mobilia et fluxibilia*) and then he moved it.[16] He is like a clockmaker who fashioned his masterpiece bit by bit, and then, once the "strings, weights" and other parts were in place (*horologium totum in suis partibus, cordis et ponderibus*), started the entire mechanism going.[17] Unlike the clockmaker, however, God did not remove himself from his masterpiece once it was set in motion. His presence is constantly necessary to keep it running smoothly. If God were to withdraw (*subtraxerit*) from nature, "we would not live, nor be moved, nor exist" (*nec vivimus, nec movemur, nec erimus*).[18]

Although God cannot withdraw from nature without it ceasing to exist, his presence in nature is not direct. He does not immediately involve himself in each and every activity that takes place in nature; he is not "the universal cause of all natural effects" (*omnium effectuum naturalium causa universalis*).[19] Instead, he exerts his activity through a chain of secondary causes that always places inferior things under the guidance of that which is superior to them (*inferiores . . . aguntur et administrantur a superioribus*).[20] Consequently, efficient causation proceeds from God through a series of secondary agents, each of which is superior to the one that immediately follows it in the order of causation. Such are the links that comprise the golden chain.

The natural secondary agents that proceed most immediately from God are the celestial movers. Efficient causation advances from God directly to the active principles that move the celestial orbs. This much was fairly well agreed upon in Henry's day.[21] What was not agreed upon, however, is the precise form taken by the active principles that move the orbs. In the course of his *Lecturae*, Henry mentions three different opinions on this subject: one held that the forms that move the celestials orbs are the same as the forms of the heavenly bodies; a second theory separated the forms of motion from the forms of the stars, arguing that motion is a quality given to them in addition to their forms; and a third view posited that the celestial movers are intellectual or rational causes that animate celestial bodies in much the same way that the human soul animates the body.[22]

Maintaining that the efficient cause of celestial motion is one with the form of the celestial body, the first opinion, is similar to maintaining that a heavy body falls down because it has the form of earth. The heavenly bodies move in a circular fashion because this is their nature—a nature that comes to them as part of their forms. This, Henry suggests, is the opinion that is held by speculative astronomers, especially those who follow Ptolemy and believe in eccentrics and epicycles.[23] The objections to this opinion are basically twofold: first, if motion comes to a celestial body with its form, then it cannot be stopped without destroying its form;[24] and second, if celestial bodies are moved by their forms and their forms remain constant, there is no way to explain how they speed up and slow down, which they obviously do.[25] Consequently, Henry rejects the notion that celestial movers are natural causes that are one with the forms of celestial bodies.

The second opinion, which requires that the celestial movers be "added natural" (*praeternaturalis*) qualities that are given to heavenly bodies in addition to their own natural forms, seems to follow a suggestion put forth by John Buridan. Buridan maintained that the same quality that is used to explain projectile motion (impetus, see chap. V, sec. 1A) can be applied to the heavens and used to account for celestial motions. He went on to argue that since there is no resistance in the heavens and since impetus is not self-wasting, the addition of impetus to heavenly bodies could account for their uninterrupted and unchanging motion.[26] Henry objects at this point that any force operating under zero resistance should produce an infinite velocity (assuming Aristotle's formula: the velocity at which an object moves is proportional to the force applied divided by the resistance of the medium through which it passes),[27] an objection that he immediately counters by noting that a specific force moves a body with a specific velocity, even in a void.[28] However, this initial support for Buridan's position is not sustained. Separating the moving quality from the form answers only the first objection raised with regard to the first opinion. This still leaves the variable velocity problem unresolved, apparently causing him to reject the second opinion too.

Having rejected the first and second opinions, Henry is left with the assumption that God's first agents are rational agents, that is, souls or what were commonly referred to as intelligences (*intelligentia*). Exactly what type of intelligences they are is another matter. He goes on to prove (with extended arguments and numerous digressions on the errors of Plato, Aristotle, and others) that the celestial movers are neither human souls nor the be stated with some assurance, that his transition from master of arts to master of theology was made without much difficulty. This and other aspects of Henry's science and theology are pleasant complements to one another.

Once the chain of cause and effect makes contact with the sublunar world, the subsequent links become considerably more complex. It is at this point—the point where the possibilities present within the seeds of matter start to become actualities through their cooperation with celestial influences—that Henry begins the process of accounting for the changes that take place in the elements. He does so by developing at some length the many paths by which the golden chain is extended past this point.

A. Alteration

The most fundamental changes that the four primary qualities cause are subsequent changes in primary qualities. When this occurs without disturbing the substantial form of the object being changed, it is called "alteration." Cold water, when influenced by a warming quality, becomes warm (accidental change), while remaining water (substantial unity).[43] Such changes can become quite complex if the subject being altered is itself complex. If, for example, the human body is conceived of as a delicately balanced mixture of primary qualities, with each organ and each part of the body having its own peculiar and proper qualitative disposition, then changes in the body's well-being, including disease and sickness, become manifestations of alteration.[44] Likewise, curing diseases becomes the process of restoring the balance of the body by inducing alteration. Sometimes this can occur naturally, as when the body produces heat in the region of a sore (*laesio*), thereby counteracting the disturbing affects of the sore.[45] Sometimes the balancing remedy comes from the outside, as is illustrated by the Biblical example of David. To counteract the chills associated with his impending death, his physicians had a young (and presumedly warm) girl brought to him to attend to his needs.[46] The complications that can enter into this curing process are many. At the very least, Henry notes, physicians must take care that the medicines they administer are not offset by the patient's own qualitative disposition.[47] Likewise, they must be careful to take stock of possible celestial influences that could counteract the potential good their medicines might bring.[48] Similar examples could be given to explain pestilences, sterilities, why human flesh is not nutritious to men, and, to raise a topic discussed above, the origin of most meteorological phenomena.[49]

B. Generation and Corruption

A slightly more complex chain of cause and effect results when the influences of the four primary qualities change not only the surface characteristics of a body (its accidents) but also the body itself (its substance). Such a change was usually referred to as "generation and corruption."[50] Changes in this instance result from the fact that the influential qualities involved are strong enough to "corrupt" the qualitative complex that characterizes the first body and "generate" a second complex that characterizes the new body.[51] Such are the changes that alchemists attempt to bring about when they mix "alum, arsenic, salammoniac, vitriol, and miscible and mineral things of this sort . . . via artificial decoctions" to produce rocks, minerals, copper, and at times even gold.[52] Similar processes also occur in nature when vapors trapped in the earth slowly ripen into minerals under the influence of the primary qualities.[53]

C. The Production of Secondary Qualities

In addition to causing accidental and substantial changes, primary qualities also give rise to secondary qualities: softness, hardness, viscosity, gravity, levity, rarity, density, odor, flavor, color, light, and so forth.[54] Although secondary qualities differ from primary qualities (they do not affect the medium through which they pass; see chap. III, sec. 2), they do extend the golden chain by causing further effects. In *Contra astrologos*, Henry gives a detailed explanation of the occult secondary qualities that cause plagues.[55] Odor, flavor, color, and so on, obviously bring in entities that have bearing on psychology, as is explained at some length in *De reductione*.[56] In addition, secondary qualities account for one of the most perplexing of all sublunar phenomena, the capacity that magnets have to bring about action (local motion) at a distance.

That primary qualities have the capacity to bring about local motion can easily be demonstrated. Heat obviously radiates over distances and brings levity to water, thereby causing it to evaporate (move up). But how is the local motion that a magnet imparts to iron explained? In responding to this question, Henry applies the same mechanism used to explain evaporation to the action of the magnet, except that the mediating quality in this case is a secondary and not a primary quality. The magnet acts on the iron by means of a "sensible active quality" (*sensibilis qualitas activa*) that elicits (*educitur*) from the iron the form of local motion. The reason that the magnet acts only on iron is that other bodies are not properly disposed to receive the peculiar secondary qualities of a magnet. And without a proper disposition, as we have already seen, influential qualities cannot cause their effects. Such an answer, Henry concludes, is at least as reasonable as assuming that an insensible occult quality is involved (*qualitas occulta aliqua insensibilis*).[57]

The introduction of alteration, generation and corruption, and secondary change into the cause-effect chain permits a great deal of flexibility in saving the phenomena of the sublunar region. However, the complexity of Henry's world view does not stop at this point. Within these causal paths he introduces, especially in *De reductione*, other factors that allow for even greater subtlety in saving the phenomena. Since both primary and secondary qualities radiate over distances, both can be reflected and refracted.[58] Moreover, since all of the changes involved are qualitative changes, each can take place by degrees through the intension and remission of forms.[59] Accordingly, as the golden chain stretches out through the sublunar region, qualities can be diverted from a direct line of action, weakened by refraction, and effective in an almost infinite degree of variations resulting from intensification and remission. Indeed, the flexibility that is ultimately built into Henry's world view is so complex that the system seems almost beyond comprehension.

However, Henry's purpose for adding reflection, refraction, intensification, and remission to an already complex world seems to have been to add precision, not confusion, to the study of cause and effect. These added factors were in his day the subject of two precise, and what he regarded to be extremely useful, sciences: perspective and the science of the latitude of forms. Although he does not say so specifically, one can well imagine that his frequent references to these sciences, in *De reductione* especially, are indicative of his firm belief that through them he could introduce into the sublunar region some of the precision so obviously present in the celestial realm. To accomplish this, he might well have had in mind carrying out the program generally described at the beginning of *De reductione*—a program that began *a posteriori* from sensible appearances (*ex sensibili apparentia*) and worked its way through estimates of the degrees of intension and remission, through the properties of primary and secondary qualities, and eventually back to the universal causes that are the subject of *philosophia communis.*[60]

The comprehensiveness and exactness with which Henry set out to describe the world in *De reductione* does not carry into his later works, with the exception of a few brief chapters added toward the end of *Contra astrologos*. When he finally turned again to scientific topics in the *Lecturae*, his conception of perspective seems more traditional, while his mentions of intension and remission are for the most part brief and not as directly under the influence of Oresme's *De configurationibus qualitatum*.[61] A parallel reading of *De reductione* and the *Lecturae* shows a marked contrast in the certainty with which descriptions are given and the confidence that Henry has in his explanations of the details associated with sublunar phenomena. Why this should be so is difficult to say. It could be that the ideas set out in *De reductione* were experimental and not really a central and permanent feature of his science. Then again, he may have discovered over the years that the high hopes he had for perspective and the study of the latitude of forms could not be realized in practice. Or it may be that the one important reservation he seems always to have had regarding the golden chain may have led him to turn to other endeavors and leave his early interests behind.

Henry's reservation regarding the golden chain concerns its exclusiveness. Throughout his major scientific writings and in the *Lecturae,* the suggestion is frequently advanced that there are two avenues by which the sublunar world is directed: one is the mediate route through the orbs as described above; the other is an immediate route via which the First Cause can act directly, although still through the orbs, in the elements. In his early work, *De habitudine*, the second, immediate route takes the form of what is described as "the influx of common natures."[62] Common natures account for unusual occurances, such as water remaining in a two-holed (top and bottom) jar if the top opening is closed. When this happens, water still retains its own proper nature, as can easily be demonstrated by uncovering the top opening and allowing the water to flow out the bottom. However, by closing the top opening, a common nature acting of itself and apart from the normal course of action (*opus . . . nec alicuius concatenationis essentialis . . . [sed] ex cursu agentis liberi*) becomes active in the water and overcomes its proper nature, thus preventing the water from flowing down.[63] Common natures also account for the fact that after rains and even before the sun shines, puddles in Rome and Paris generate the same type of worms; and for such curious phenomena as a group of frogs attacking a man who had earlier killed some among them.[64]

Although Henry does not clearly describe how common natures act, it is obvious that they are similar in action to supernatural causes. Since common natures are a result of a free agent, and since intelligences acting through the orbs have no free wills, the influx must come from a prior, and presumedly the First, cause.[65] There is, therefore, little difference between the action of common natures and supernatural causes as described in *Quaestiones super perspectivam, Tractatus de discretione spirituum,* or the *Lecturae.*[66] It seems likely, in this light, that Henry consistently believed that any effect in this world might stem from causes that lie outside the golden chain. Such a belief does not, however, dispense with or invalidate the golden chain. The world still functions mostly in a natural way. Supernatural causation accounts for the unusual and spiritual, not the normal and natural. It is therefore wrong to suggest, as one scholar has, that "In place of the standard hypothesis of astrology that God committed the control of the world of inferior natures to the stars, Henry substitutes a theory of the immanence of the First Cause throughout the universe."[67] If this were all there were to Henry's conception of cause and effect, he could then have dispensed with the science that made such a scandalous use of physics, i.e. astrology, with one simple, fideistic blow–something he certainly did not do.

2. Astrology and Its Refutation

Astrology is the science that seeks to determine by observing the stars the events that will subsequently transpire in the elements. It is a predictive science whose efficacy, according to Henry, comes last in the order of certainty after medicine and meteorology.[68] In order to justify such an endeavor, it is necessary to believe, if not demonstrate, that the heavens and earth are somehow intimately connected. The stars must in some way be a mirror of what goes on below, so that one can observe in that mirror sublunar events. Obviously, if God does not operate through the stars, then nature depends on God not the stars, and astrology becomes a useless endeavor. An "immanent first cause" directly "operating throughout the universe" would indeed destroy "the standard hypotheses of astrology." This is not, however, how Henry proceeds in his arguments against astrology, as the preceding discussion of intelligences and influences should have suggested.

By severely limiting the role the orbs play in directing inferior events and by assigning the majority of the causes of earthly phenomena to the elements, Henry makes the orbs such a minor part of the causal process that nothing specific can be predicted from their motions alone. Therefore, rather than directing all of their attention to the heavens, astrologers should, according to Henry, spend more time studying the elements. Indeed, if one considers that in order to determine whether a particular influence will be received by and effect change in the elements (recalling that reception is a cooperative process) it is necessary to take into consideration: (1) changes in the qualities of the earth over time (Egypt was once a wet land, now it is dry), (2) seasonal variations, (3) changes in the locations of the stars (which changes affect the angle at which their rays are received), and (4) intermediary obstacles such as clouds, it is no wonder that Henry remarks in disgust:

> It is amazing how certain persons who are still beginners (*beani*) in the art of astrology attempt rashly to meddle with astrological judgments [and] with their lies [provoke] the minds of the people to nonsensical acts by parading before [them their] vanities.[69]

Clearly, astrology cannot have any validity if it is not predicated on a careful and complete study of the elements. Simply observing the stars is not sufficient.

Such an objection, of course, leaves the door open to astrology. It does not deny stellar influences *per se*; it only doubts their knowability. But this presents no problem as far as Henry is concerned. He is not against astrology *per se*, only the errors of astrology and its practitioners. These comprise the focal point of his attack–an attack that begins in *Contra astrologos* with the stark pronouncement: "The University of Paris detests following superfluous vanities!"[70] The vanities in this case include making predictions about the mortality of men, wars, disease-producing winds, and the other events that astrologers attempt to relate to conjunctions. It is error and superstition that again and again prompt him to turn from the proper subject of his text in the *Lecturae* and proceed with yet one more tirade against astrology. Sometimes these tirades are put forth with some reluctance. But the added consideration of protecting the innocent, even "certain Christians" who follow the authority of the ancients "like horses and mules, which have no intellect," overcomes this reluctance and allows him to proceed.[71] For example, he argues that

> although the sayings of that vain discipline should not be dignified with a serious response, nevertheless, for certain of the more simple-minded who happen to adhere too swiftly to the authorities of such superstitions, by what titillating itch of vanity and lightness I do not know, I wish to say something against the crude and undistinguished zealotry of astrological superstition.[72]

The well-orchestrated attacks that follow such pronouncements touch upon virtually every aspect of astrological prediction. Whereas lesser issues are sometimes passed over "for the sake of brevity,"[73] when it comes to the vanities and errors of the age, no point seems too insignificant to warrant at least a few words. The details of these attacks quite obviously lie outside the scope of this book. An entire work could, and indeed should, be written on Henry's views on astrology. For the present, a summary of his major areas of concern must suffice.

(a) *Technical Astrology*. Even though Henry is firmly opposed to assigning specific influences to specific stars (apart from the four qualities), he spends a great deal of time showing how even if such influences existed, astrologers cannot, or at least do not, know them. His objections in this case run from simply stating that the remoteness of the heavens does not permit knowledge about them—we do not even know the order of the planets[74]—to fairly sophisticated rejections of the astronomical system that underlies astrological prediction, Ptolemaic astronomy. The latter objection obviously rests heavily on his distrust of eccentrics and epicycles.[75]

But even if eccentrics and epicycles could be accepted, Henry would still have several objections: the tables drawn up by Ptolemy were already in error at the time of King Alfonso the Wise (1252–1284); a simple check of the actual position of Mars versus its calculated position will demonstrate that Ptolemy's equations are not accurate; since stars do not effect one another, giving special influences to conjunctions, eclipses, and various other stellar patterns is useless; and even if stars effect one another, their constantly changing pattern would make it impossible to know all of the variations that occur.[76] Consequently, even if his arguments about limited influences are

Henry very much believes, as did Aristotle, in a world that is drawn together by precise cause-effect relationships that operate according to the laws of nature, God's laws that were ordained through creation. Therefore, with the exception of unusual occurrences, what happens below is unalterably and mechanically linked (that is, operating like a machine with each part moving the next in a designated way) to celestial events. Henry does not deny this fundamental assumption, which is so crucial to astrology. However, he is also fully aware that the complexities of this mechanical process are so great that it is extremely difficult, if not impossible, for humans to ascend the chain and understand cause at its source. It is this fact, more than any other, that undermines the science of astrology and makes it, as it was ordinarily practiced, a useless endeavor. This would, of course, not be so if God had, as he ornamented the elements through the golden chain, created a simple world. However, he did not, as the events of the last two days of creation make clear.

CHAPTER VII

Day Five: Plants, Animals, and the Biological Sciences

And God said, "let the waters bring forth of the living spirit the reptile upon the earth and the bird beneath the firmament of the heaven." And God created great sea monsters and each living and moving creature that the waters had brought forth in their kind, and each bird according to its kind. And God saw that it was good. And he blessed them saying, "thrive and multiply, and fill the waters of the sea, and let the birds be multiplied above the earth." And the evening and morning were made, day five. Genesis 1:20–23.

Once the details of the celestial region had been filled in, God turned his attention to the elements that lie below. Day three saw the latter disposed throughout the sublunar world, with each either in, or striving for, its proper place. The work of the fifth and sixth days adds to the elemental spheres their final details, the forms of life that were commonly thought to properly pertain to each. The waters bring forth all forms of marine life, including reptiles and sea monsters. Air becomes the natural abode of birds. Lastly, on the sixth day, earth gives rise to animals and the human species. Each living creature is in turn given a command that is responsible for the uniqueness of life: "thrive and multiply." Growth and reproduction are what distinguish the animate from the inanimate and permit living creatures to populate the earth.

Of the various living creatures brought forth from the elements, there can be little doubt that the one which had the most fascination for and engendered the most discussion among medieval schoolmen was the human animal. Whereas volume upon volume was devoted to philosophical, theological, and empirical speculations about the nature of man, the remainder of the biological sciences remained, throughout the Middle Ages, close to the level set by Pliny and the other encyclopedists.

Henry is no exception in this regard. His discussions of other forms of life are not extensive in comparison to the many lectures prompted by the creation of man. When drawn together, however, the information on plants and land animals set out in conjunction with the work of the third and sixth days plus the information on birds and fishes included in the work of the fifth day, provides a comprehensive insight into how the scholastic scientist visualized the world of living creatures surrounding him. In drawing this information together in the present chapter, I have somewhat arbitrarily pursued the discussion on two levels: first, on a more theoretical level, as life can be considered in terms of physiology and classification (sec. 1); and second, on a more popular level, as creatures exist or were thought to exist in the world (sec. 2).

1. Physiology and Classification

If physiology is assumed to be the study of the manner in which living organisms function, then medieval physiology can be defined as the study of the manifestations of the soul's actions in the body. As with everything else in the medieval world, living creatures must be thought of in terms of matter and form. To be alive means to have the form of life, the soul, impressed on and active in the matter that makes up the body. "The soul is the act of an organic physical body having life in potency."[1] The net result of this conception is that treatises entitled "on the soul," after Aristotle's *De anima*, are to some extent physiological treatises. Moreover, such treatises comprise a major source for the study of scholastic conceptions of physiology, since what we would consider more properly physiological treatises–such as Aristotle's *Historia animalium*, *De partibus animalium*, and *De motu animalium*, as well as other more purely physiological works in the medical tradition of Galen (second century a.d.)–had fairly limited circulation in scholastic circles.[2] In this regard, Henry is no exception. He makes only limited use of Albert's extensive writings on animals, cites Avicenna's prestigious *Canon* on medicine only once, and never quotes Galen by name. His only extended discussions outside the *Lecturae* of what it means to be alive are contained in his *Egregia puncta et notata de anima*.

The soul, or active form of the living body, manifests itself in a number of ways: living things grow and reproduce; some have the capacity to sense and respond in quasi-intelligent ways to stimuli; all, excluding plants and some lower animals, can move; and one special form of life, the human form, has the capacity to think. It is obvious, therefore, that even though each living creature has only one form, this form activates and controls a number of very specific and distinct operations. (1) Most fundamentally, to be alive means to have the power, called the vegetative power (*potentia vegetativa*), to "thrive and multiply." This encompasses securing nutrition, utilizing that nutrition for growth, and reproducing. Thereafter, to be alive can mean (2) to possess the capacity to sense, called the sensitive power (*potentia sensitiva*), and (3) to react to what is sensed by means of desire or appetite (*potentia appetitiva*). Reacting involves local motion, which requires (4) a motive power (*potentia secundum locum motiva*). Finally, if life is extended to include the human species, one last power becomes important–(5) the rational or intellective power (*potentia intellectiva*), which clearly sets the human being apart from all other living creatures.[3]

The human soul, as the most noble of all souls, has within the range of its capacities all of the above powers. Such is not the case, however, with the souls of other living creatures. Plants do not have the ability to move; their vital functions are limited to growth and reproduction. Flies move but have no memories; they will return to the same spot where only a moment before an attempt was made to swat them.[4] Higher animals have memories and other senses, but they cannot think; spiders and bees produce webs and hives by instinct (*naturalis instinctus*) and not by reason.[5]

It is evident, therefore, that different types of creatures have different types of souls. This provides, as most biologists from Aristotle on fully realized, a very convenient way to classify them. Those with the least powers–plants–simply grow and reproduce. Those with more complex powers–animals–grow, reproduce, sense, and move. (Henry divides this group into two subgroups: animals with relatively simple sensory capacities, such as shellfish [*conchilia*], and those that have better developed sensory capacities.) Those with the most powers–humans–grow, reproduce, sense, move, and think.[6]

From the soul, therefore, attention is naturally directed to function, and from function to organization–that is, from living form to vital processes, to the hierarchical arrangement of living creatures. This sets the general outline of what it means to be alive. Hereafter, as theologian turned biologist, Henry goes on to fill in the details.

A. Physiology

No attempt is made in the *Lecturae* to explain systematically how living creatures function. For the most part, Henry's interests focus on reproduction, with growth and the other vital processes being mentioned only in passing. Whether he knew more about physiology than was set out in the *Lecturae* is impossible to determine. What can be said is that Henry's students, after sitting through the lectures on the six days, would have had fairly limited information about the particular actions that the soul has in the bodies of plants and animals.

Plants reproduce by means of seeds that depend on the elements and not on a surrounding egg for their nutrition. When seed and soil come together, the seed germinates, partially because of the nature of the seed itself and partially because of cold, dry, and other accidental qualities that are in the earth (*per . . . frigiditatem et siccitatem et per qualitates alias quae accidentaliter in sunt ei*).[7] Thereupon, the germinated seed grows with the added help of celestial influences by attracting nutriments and converting these through a "digestive power" (*per virtutem digestivam*) into foliage, flowers, fruits, and more seeds.[8] This results in a cycle of seed to plant to seed, which accounts for the multiplication and propagation of plants. Since this cycle depends as much on the disposition of the qualities of the earth and celestial influences as it does on the seed itself, it is possible to suggest that a change in the disposition of the earth might change the cycle. For example, winter wheat (*siligo*) after many plantings might turn into hard wheat (*triticum*). If it does (Henry seems to doubt that this example, which is reported by Pliny, is true), then he maintains that the change is accidental (*quadam differentia naturali accidentali*) and not substantial.[9] He is therefore not readily disposed, in this case at least, toward accepting the mutability of species. Plants produce plants of the same species.

The manner in which plants grow and reproduce is closely paralleled by the manner in which frogs, bees, some types of fishes, and the worms that arise from spontaneous generation propagate themselves. Frogs, bees, and fish spread their seeds among the elements, where they eventually begin to grow into mature adults. Worms, especially the worms of putrefaction, arise from occult or hidden seeds (*seminales materiae occultae*) that reside in the elements and germinate under special celestial or supernatural influences.[10] Hereafter, each presumedly continues to grow by incorporating nutriments into its structure by means of the digestive power.

Accounting for the origin of animals as well as plants with seeds germinating in the earth is not without precedent. The doctrine that Henry is taking over in this instance and applying to a specific segment of nature is the more general doctrine of "seminal reasons" used by Augustine to account for the origin of the entire design of creation (chap. II, sec. 2). Just as God originally gives to the matter of the universe the potential to cooperate with the activity of forms in shaping the universe, so too he gives to the matter out of which animals will be formed the potential as seeds (*seminales*) to become animals. How these seeds function is set out in some detail in the *Lecturae*.[11]

There are three types of *seminales* that take part in production of animals. Those most directly (*propinquae*) involved in the procreative process are the seeds (*seminales transeuntes*) that actually pass from the initiator of procreation to the site of procreation. Such would be the seeds (*grana*) of plants and the sperm (*seminales*) of animals. These seeds act in cooperation with other seeds (*rationes seminales,* which are *remotae*) that are in matter and dispose that matter toward the reception of specific forms. Together, these two seeds are all that are normally needed to produce progeny. The active seeds (*seminales transeuntes*) cooperate with the passive seeds (*rationes seminales*) to produce progeny. However, in special instances, as *in principio*, there is a third type of seed in matter (*seminales obedentiales,* which are *remotissimae*) that can cooperate directly with God and produce individuals apart from the natural course of events. Since the latter are involved in supernatural causation, they seldom are relevant to a consideration of how nature functions.[12]

The doctrine of *seminales* has one very interesting biological consequence. Since the process of *seminales* becoming animals or plants involves mutual interaction, changes in the material conditions under which the generative process takes place can lead to the production of different types of creatures. Thus, rotting cow flesh gives rise to bees, horse flesh gives rise to beetles, mule flesh gives rise to locusts, and rotting crabs give rise to scorpions.[13] In these instances, it is as much the material substratum of reproduction that determines what the progeny will be as the active seeds that initiate reproduction. Moreover, since in many animals this process takes place at the mercy of the elements, changes in the disposition of the elements can drastically alter the nature of the progeny that are produced. Indeed, this may well be the origin of sea monsters. Since fish distribute their seeds in water, a slight turbulance in the water in which the seeds are distributed combined with the normal distortion of celestial influences that results from their passing through water, could easily pervert the original course of generation and give rise to monsters instead of fish.[14] However, according to Henry, such chance generation does not happen in higher creatures unless God steps in. Thus, just as he doubts that winter wheat naturally mutates to hard wheat, he does not entertain the possibility of a horse giving rise to a cow. Among higher and more perfect animals, species are fixed. Nevertheless, he is perfectly willing to admit that within the ranks of lesser, imperfect creatures species may not be fixed.

What results from the actions of *seminales* are, of course, mature living creatures. The seeds provide the mechanism through which

> God, the creator, wishing there to be as many different types of things as were needed for the proper ordering of the universe and his glory . . . , established in living creatures (*eis*) a marvelous array and diversity of parts and members, both interior and exterior, as well as a varied and wonderful placement of organs.[15]

This "marvelous array" includes the means to seek out food, such as roots in plants and the bony structures of animals. It includes the internal organs by which food is digested and thereafter distributed throughout or, if it is in excess (*superfluitas*), expelled from the body. Most importantly, it includes the one organ that mirrors in the microcosm God's role in the macrocosm, the heart. Just as God rules the worldly machine (*machina mundi*), so too the body is ruled and enlivened by the heat of the heart, for without its action, we would die.[16]

Henry never says exactly how this marvelous array of organs functions. He is certain that they do function in the best possible way and that each part of the living creature is so placed and so ordered that it serves its proper function and preserves the integrity of the creature. "Coats, fur, horns, teeth, claws, scales, and other forms of armor," serve to protect animals.[17] The webbed feet of water birds allow them to get food. And the speed of the rabbit, which is a function of its structure, helps it flee from predators. But aside from the assurance that each part, however it functions, has a purpose, Henry adds very little. For the listener who wants to know

> . . . more in particular about the wonderful manner in which animals and their parts are organized in order to make manifest divine power and wisdom, the twenty-six books that Albert the Great wrote on animals should be read.[18]

At least if he did not himself supply the details regarding physiology, he sent his students to what was probably the best text on the world of living creatures readily available at the time.

B. Classification

All living creatures can be divided into the general categories of plants, animals, and humans. If one attempts to further clarify how they differ from one another, many ordering principles become apparent. Most simply, it can be argued, following the Genesis account of creation, that God assigned one group of creatures to each of the elements: land animals to earth, fish to water, birds to air, and demons, angels, or some other form of life to fire. Each animal would in turn have the nobility of its own element and depend physiologically on that element for its existence.[19]

Henry rejects this correlation for a number of reasons. In the first place, he doubts that any animal lives in the fire that is immediately below the sphere of the moon. This leads him to interpret Augustine's pronouncement about animals living in fire as referring to terrestrial fire.[20] Secondly, he doubts that any animal could live on one element, such as the fish "allec" (*allex*) on water, the mole (*talpa*) on earth, the salamander (*salamandra*) on fire, and the chameleon (*chamaeleon*) on air.[21] Thirdly, living creatures are a mixture of the four elements with earth predominating in each, since they are all heavy and opaque.[22] Finally, the elemental composition that a particular animal has does not always correspond to its degree of perfection. The baseness of the element earth, for example, which produces melancholic (*melancholicus*) temperaments docs not dictate that persons with such temperaments should be classified as inferior. In fact, just the opposite is true in this case. As Aristotle had noted, illustrious men and the founders of the sciences were mostly melancholic, and they were clearly no less noble than other men.[23]

Having rejected this relatively simple scheme of classification, Henry advances other ways of setting one creature apart from another. Each reflects his thoughts on a particular aspect of the living world, or more simply, on how one aspect of the soul's operation manifests itself in different creatures. None of them is designed to provide a comprehensive system for classifying all animals, and is therefore not expanded to include a complete analysis of many forms of life. What follows is given more by way of examples in the *Lecturae* than as intended discussions on classification.

(1) Motion. As has already been noted in conjunction with the elements (chap. V, sec. 1), motion, taken in its most general sense, applies to any change of form. Accordingly, it is as correct to say that a plant moves when it grows (generation or augmentation) as it is to say that a bird moves when it flies through the air (local motion). Working from this general definition of motion, Henry goes on to present an analysis of how living creatures move.

Some living creatures simply increase in size (plants), while others move locally (animals). Those that move locally move only their parts (*solum secundum partem,* e.g., a sponge) or their entire body (*secundum totum,* progressive motion). Those that move their entire body do so either with feet (men and quadrupeds) or without feet. The latter include winged creatures of two types, those with feathered wings (birds) and those with membranous wings (bats and flies).[24]

(2) Reproduction. At one point in *Historia animalium*, Aristotle noted that animals reproduce in three ways: by bearing live progeny (viviparous), by means of eggs (oviparous), and by producing worm-like progeny that grow into mature adults (vermiparous).[25] Similarly, Henry suggests that animals produce their young in one of two ways: either within the body (humans and the other "perfect animals") or outside the body (*extra ventrem*). Those that produce their young outside the body do so either by means of eggs (Aristotle's oviparous animals) or by means of seeds (Aristotle's vermiparous animals) that germinate into worm-like creatures in much the same way that seeds grow into plants.[26] The latter group includes bees, ants, frogs, and some types of fishes–all of which produce progeny that are similar to themselves–as well as the diverse worms that result from putrefaction and that may or may not ultimately resemble the prior generation.[27]

(3) Sensation. Since plants have no animal soul, they cannot sense and hence can easily be distinguished from animals. Animals do not all have the same complement of senses and thus can be further distinguished from one another. Birds, for example, have a full complement of senses: touch, taste, smell, hearing, and vision. Shellfish and other imperfect animals have only

touch. Therefore, birds are superior to shellfish. Caution must be used in reaching conclusions about perfection from the senses, however. Just because bees, ants, and spiders produce things by their labors (that is, they seemingly have intellectual as well as sensory capacities) does not mean that they are superior to the lion, bull, and horse. The labors of these tiny animals, as was mentioned, stem from instinct and therefore are not indicative of superior sensory capacities.[28]

(4) Vital organs. Just as an evaluation of sensation is ultimately reduced to an appraisal of the number of sense organs that animals have, so too, an evaluation of physiological functions can be reduced to the number of vital organs that each has, thereby providing another basis for classification. Henry does not pursue this method, other than to note that ultimately each animal must be compared to man, since man has a complete complement of vital organs: brain, heart, liver, genitals, and so on.[29] Again, caution needs to be used in drawing conclusions from this information. The crab, for example, appears to be filled with a homogeneous fluid (*humore vel humoribus homogeniis*) and, therefore, would seem to be extremely primitive. However, its lack of internal differentiation is more than compensated for by its external organs, and thus it is more perfect than might otherwise be expected.[30]

The possible ways of classifying living creatures do not end with these four (or five, if the rejected correlation to the elements is included). Animals, in particular, are very complex. Any aspect of their living being could be used to distinguish one from another: the noises that they make, the services that they perform for man, the products of their industry, their cleverness or stupidity, what shape they are, how their organs (both internal and external) are disposed, and so on.[31] Moreover, each aspect could be evaluated in a number of ways. One might judge the placement of parts within animals on the basis of how they fit together with one another and function like the parts of a clock (*pars horologii*), or how they aid in procuring food, or how they permit an animal to serve man, or how they relate to the elements, or, most generally, in terms of how they fit into God's overall design for the universe.[32] In short, classification, along with the evaluation of the physiological activities that are related to the various schemes for classification, is a very open-ended project for Henry. There is no *one* way to classify living creatures; there are many. Henry does not pursue classification as a topic in order to find out what makes a horse different from a fly; the difference between them is obvious. His reasons for pursuing classification are entirely different.

What makes classification important, and the reason that some time is spent discussing it in the *Lecturae*, is that it aids in understanding the plan of God's creation. The general outline and course of the work of the six days is clear. God began with nothing and produced a finished masterpiece. He proceeded from the imperfect to the perfect–that is, from plants to animals and eventually to the most complete of all creatures, man.[33] In accordance with this order, it is also clear that all creation, and especially the living portion, is arranged in hierarchical fashion. The pieces are sequentially set out to ornament the world, each of which is better than the one that comes before. There is a "great chain of being."[34]

What is not clear, however, and this is where classification becomes important, is exactly how the chain or hierarchy is ordered. Unlike the world of elements, the world of living creatures does not operate according to a fixed causal order. There is no "golden chain" of cause and effect that extends from man to mold. Men do not create elephants, elephants mice, and so on down the chain, and so there is no simple way to explain why the human animal is the most superior of all living creatures or why a lion is superior to an ant. Classification attempts to demonstrate that there is order within the world of living creatures, an order that extends in its own noncausal great chain of being from worms, plants, and shellfish to horses, dogs, lions, and men.

In demonstrating that there is order, Henry does not prove his assertions. In contrast to his discussion of the golden chain, he does not attempt to explain the cause and operation, the teleology and ontology, of the great chain of being.[35] At this point in the *Lecturae*, his approach to nature becomes more that of a museum-goer than a scientist. He comes to the gallery marked "living creatures" to be inspired, not to seek precise knowledge about the Creator. How and why these creatures came to be here is obvious; one simply has to read the sign over the door of the gallery: "These are God's creatures set out as part of creation." Knowing and believing this, he simply sets to work describing to his fellow museum-goers the marvels that are found in the gallery. This permits him to move from creature to creature, briefly pointing out the merits of each as it further exemplifies the ingenuity of the Master Artisan and without digging more deeply into the metaphysical and theological problems that are inherent in accounting for this ingenuity. In the process he does manage to stop at a number of different habitats and briefly describe in more popular and less functional terms the creatures that are found therein.

2. Descriptive Biology

The guidebook upon which Henry relies most for his tour through the gallery of living creatures is undoubtedly a work he had himself recommended to his students, Albert's *De animalibus*.[36] The wealth of information, sometimes firsthand, contained in this work makes it an ideal source for learning about not only the anatomy and physiology of animals but also their gross characteristics and habits (the study of which I have referred to as *descriptive biology*). Henry derived much of his general knowledge about animals from this work. However, like so many of his contemporaries, he also made extensive use of several more popular sources, including Isidore's *Etymologiae*, Ambrose's *Hexaemeron*, and a source used by both Isidore and Ambrose, Pliny's *Naturalis historiae*. As a result, the descriptive biology set out in the *Lecturae* represents a mixture of reasonably accurate information combined with popular beliefs and misconceptions. The latter could be ignored were it not for the fact that Henry, in common with most of his contemporaries, accepts many of the popular elements as "facts" and uses them as data in accounting for the phenomena of nature.

One example drawn from the plant kingdom should serve to illustrate this last point. Henry accepts the common belief that garlic (*allium*) has the power to impede the action of a magnet.[37] The significance of this seemingly trivial "fact" is great. The magnet provides one obvious and readily accepted example of occult actions in the sublunar world. As such, its operation is similar (and thus presents an analogous case that can be used for comparison) to other occult phenomena such as the action of celestial influences or the operation of the senses.[38] If something as simple as garlic can impede the action of a magnet, then there is no reason why more complex powers in the elements cannot impede the actions of celestial influences, which is exactly what Henry argues. Having the certain knowledge that garlic can impede the action of a magnet, he is confident that celestial influences do not have an absolute effect on the elements and can be altered by elemental forces, a confidence that adds one more piece of evidence to his case against astrology.[39] The obvious consequence of the use of such "facts" in scientific discussions is that it is impossible to exclude from a description of a medieval schoolman's conception of plants and animals even the most insignificant and incredible information. Descriptive biology includes as much or more of the irrelevant, by our standards, as the relevant. This will become obvious as we take up the details of descriptive biology as set out in the *Lecturae*, beginning with a description of a number of plants and then proceeding up the chain of being to the lower and finally higher and more perfect animals.

Among the plants that are mentioned in the *Lecturae*, cress (*nasturtium*) is cited as an example of one whose nature is quite the opposite of the conditions under which it grows. It has a hot-dry nature—Pliny notes that it derives its name from the pain that it gives to the nostrils ("nostril-tormenter," from *naris* and *torqueo*)—even though it grows in water.[40] The violet (*violis*), lily (*lilium*), and rose (*rosa*) are noted for their brilliant colors; their purples, whiteness, and reds give delights comparable to the delights received from the Saints.[41] Rhubarb (*rheubarbarum*) causes jaundice (*cholera*) as naturally as magnets attract iron and rocks fall down.[42] A similar natural action is given to pepper (*piper*) and ginger (*zingiberi*), which Henry says produce the primary quality of heat directly and not some occult quality that in turn gives off heat.[43] Wheat and peas are mentioned for their reproductive capacities. Wheat (*triticum vel siligo*) produces as many as sixty grains, whereas in peas (*pisum*) there is sometimes found one or two hundred seeds.[44] All of which leads Henry, following Ambrose, to note that Solomon in all of his wisdom may have been able to describe all sorts of plants, "from the cedar that is in Lebanon to the hyssop that grows out of a wall" (1 Kings 4:33), but still he could not have known the numerous individual properties of each.[45]

Turning to animals, several types have water as their natural home, including fish, sea monsters, as well as some reptiles and amphibians. Fish, as has already been noted, reproduce by spreading their seeds throughout the water in which they live. Henry notes that this process is extremely productive, since a single fish through one birth (*una generatione*) can produce as many as one thousand progeny.[46] The unlimited growth of fish that would naturally result from such prolific generation is kept in check by predators, such as the sea monster, which in turn is prevented from destroying all fish by its large size. Small fish instinctively (*naturali instinctu*) flee to shallow portions of the ocean (*ripae mare*) where they are safe.[47] Henry thus seems to believe there is a natural balance in nature that prevents animals from either becoming extinct or overrunning the earth.

A second form of life found in water, the sea monster (*cetus*), is, according to Henry's reading of Pliny, the largest animal that exists. One sea monster reportedly would occupy as much as four acres (*jugeri*) of land.[48] The sea monster is known by many names, including whale (*balaena*), dragon (*draco*), leviathan (*leviathan*), and sea monster (*belva marina*).[49] Its immense size led some persons to suggest that if they were to multiply, they would destroy the world (*saeculum*), make the seas unnavigable, and devour all fish. Henry replies, however, that once they were created, God made them infertile and hence reserved their terror for the coming age (*in futuro saeculo*).[50]

A third group of animals that lives in water, and in this case on land too, is that which encompasses reptiles. In setting this group apart from other animals, Henry ignores the fact that all are egg-bearing and focuses instead on their manner of progressive motion. No reptile, including those with feet, moves in exactly the same way as men and quadrupeds. Newts (*stelliones*) and lizards (*lacertae*) have feet that are used for only brief periods of time; serpents have no feet and move by using their body (*vi costarum*); worms draw themselves along (*se trahunt*); and fish, which Henry in this case includes under reptiles, use their fins and tails to propel themselves through the water.[51] Even the turtle (*turtus*) and crocodile (*crocodilus*) can be classified as reptiles, since when they walk they keep their legs and hips close to the ground, unlike other quadrupeds.[52]

Serpents can be distinguished from other reptiles in their ability to raise the forepart of their bodies in the air when they want to; reptiles cannot.[53] From among the sixty or so species of serpents that had been discussed by Albert, Henry draws several examples, primarily to set forth appropriate moral lessons. For instance, the bite of the asp (*aspis*) causes senselessness, chills, and death by a deep sleep in about three hours. The asp is also known for its ability to protect things put into its custody. If it swallows a precious stone that can only be retrieved by a special song (*incantatio*), it will place one ear against the ground and plug the other with its tail so that it does not hear the special song.[54] The viper cerastes (*cerastes*) has eight horns on its head like the horns of a sheep; it is the color of dust and comes from Africa. It had been reported (although Henry notes following Albert that the reports have not been proven [*non est satis probatum*]) that the horns of this snake sweat in the presence of poisons and therefore are used to make the handles of small table knives (*manubria cultellorum*) that nobles and bishops (Albert says simply nobles) use.[55] Similar descriptions and attributes are related for the *dypsas, draco, jaculum, natrix,* and *cuphestus* (*cafezatus*).[56]

One further group of animals is also found in water but lives equally as well on land, the group known as amphibians. Henry mentions two examples in this group, the crocodile (*crocodilus*), which had already been classified as a reptile, and hippopotamus (*hippopotamus*). The crocodile, which gets its name from its saffron-like color (*croceus*), is a large animal–more than 20 cubits long according to Isidore–that swims in water and walks on four legs on land. It has immense jaws, of which only the top one moves, numerous teeth, and a heavy armored covering (*armatum durissimam superius cutem habens*). The hippopotamus is also called the river horse and cow of the river. Although Henry does not repeat this story, it was commonly believed that the hippopotamus would back into fields upon which it was feeding, thereby making it seem as though it had left the field so that traps would not be set to catch it.[57]

Animals that have wings and move through the air are broadly classified as birds. What types of wings they have makes no difference; those with membranous wings (flies and bats) as well as those with feathered wings are considered to be birds.[58] Structurally, birds are fairly unique. Unlike other animals, they are slim and graceful toward the rear and heavy toward the front. This aids them in flight.[59] Since they eat constantly and must move about to get food, they are oviparous. A pregnant bird would not be able to fly.[60] Their light nature makes them talkative. This is proven by the fact that lighter and smaller birds chatter more than heavier birds.[61] Many examples of laudable and damnable behavior can be found among birds. Cranes (*grues*), for example, provide an excellent example of an efficient military organization; their night-guards voluntarily pursue their rounds and then, before retiring, wake the next guards to take their places.[62] Falcons (*falcones*) and other predators (*aves rapidae*) are similar to tyrants in that they oppress and live off of society. Consequently, they are expelled from society and live solitary existences on "the highest mountains and rocky cliffs" (*in altissimis montibus et rupibus*).[63]

Finally, there are the numerous varieties of land creatures, from wild and savage beasts, such as the lion (*leo*), bear (*ursus*), and wolf (*lupus*), to the various types of domestic animals. The largest among this group is the elephant (*elephantus*). Its trunk, which it uses in place of a hand in war and eating, measures as long as ten cubits (ca. fifteen feet). The elephant can carry from twelve to as many as forty men in a wooden structure (*castrum ligneum*) that is placed on its back. For this reason it is frequently used in war. Elephants are also moved by the sound of the human voice and bow when in the presence of a king.[64] As with birds, land animals provide numerous moral examples to follow. Elephants, dogs, and horses are well known for their faithfulness. The service of the lion to Jerome is, of course, almost without equal.[65] The actions of animals also provide examples of fundamental psychological processes, as a sheep's instinctive fear of wolves. This latter example was universally held to be true. Henry notes that Albert had never seen (*non sum expertus*) such fear in sheep, although he had experienced the natural hostility between geese (*anseres*) and eagles (*acquilae*).[66] Such is the tour that Henry gives his students through the gallery of living creatures.

3. Biology and Superstition

As was noted, Henry's main objective in presenting this potpourri of descriptive detail as well as the speculations about physiology and classification is to demonstrate the marvels of creation. Wherever he looks in nature he finds evidence of the Master Artisan. When the plan of the Creator is not obvious, he seeks it out. It is not enough for him to know what a sea monster is. He has to know why this largest of all animals lives in water and not on land, what its purpose is in nature, whether just as there is a largest animal there is a smallest animal, and where, if there is a smallest animal, it would live.[67] In this limited sense, his investigation of living creatures is motivated by teleology. First cause or purpose is always on his mind.

At times this search for purpose leads him to adopt views of nature that are quite similar to our own. For example, it can be inferred from several brief passages in the *Lecturae* that he considers the world of living creatures to be an interdependent community, just as might be suggested by someone who is familiar with modern ecological thought. Individuals may exist as individuals, but they also must participate in and help preserve nature. It is for this reason that there are natural predators in nature, "the cat toward the mouse, the spider toward the fly, and dogs toward wolves."[68] Moreover, animals help keep balance and order within the elements; frogs and toads purify the air of venomous spirits and help keep it clean.[69] When the balance of nature is upset, perhaps by the actions of men, the world of living creatures suffers. The Fall of man turned the peaceful animals of paradise into scavengers and ferocious beasts. More recently, civil strife in Italy at the time of Augustine turned domestic animals loose and changed them into wild and untamed creatures.[70] In sum, just as each organism runs and fits together like a small clock, with each part depending on every other part, so too, the larger world of living creatures maintains a semblance of order, with each part carrying out its own designated role. Even before modern ecological thought, Henry seems to have realized that the living world that surrounded him was a delicately balanced community of interdependent beings that needed one another to maintain that balance.

More commonly, however, the search for purpose in nature leads Henry to adopt superstitions and old wives's tales that turn what might otherwise be considered biology into a futile endeavor to explain away the complexities of living beings. Such a judgment would, of course, have surprised Henry, since he rejected superstition with the same relentless spirit that he turned against astrology. In doing so, however, he does not reject the incredible stories put forth by Pliny, Isidore, and Albert; but rather, he rejects something very specific that lies outside scholarly circles.

Superstition for Henry consists of popular beliefs held by "rustics," "poets," "farmers," and other purveyors of vain traditions. Superstition includes the belief in domestic spirits, as maintained by those who live in Norway (*Norbergia*), the East European lands (*Dacia*) bordering on the Danube, and other outlying regions.[71] It includes such widely held opinions as: the worm of an oak tree that is in fruit will bring fertility (*si in pomo quercus inveniatur terma iudicant fertilitatem*).[72] Certainly, nothing that he reports would, in his own mind, fall into this category. Most, if not all, of the ideas that are set forth in the *Lecturae* are based on established authority, not popular belief.

In addition, superstition involves prognostication. Purveyors of superstition endeavor to do more than relate facts; they attempt to predict. Since these predictions always, in one way or another lead to the consideration of stellar influences, Henry rejects this aspect of superstition as well, including prognostications from earth (geomancy), from water (hydromancy), from fire (pyromancy), from birds (augery), from beasts (bestomancy), and by communing with dead spirits (necromancy).[73] None of the descriptions he relates are directed toward predictions. At most, they lead to moral lessons. So here too his beliefs fall outside the pale of superstition. From Henry's vantage point, what is set out as biological "fact" and what falls under the domain of superstition have nothing in common.

From our vantage point, however, they do have much in common. The refuge of authority and tradition does not preserve Henry from the numerous erroneous and incredible beliefs (by our standards) that became incorporated into his science. And one wonders, as indeed Lynn Thorndike did, why someone so committed to finding fault with others did not turn his critical powers toward his own work. To paraphrase Thorndike's criticisms: how could he have spent so much time pointing out the numerous, unfounded beliefs held by astrologers, while basing much of what he had to say on "the equally incorrect scientific hypotheses as spontaneous generation . . . or the explanation of comets as formed from exhalations?"[74] Should we not expect more of Henry, vis-à-vis superstition, if his approach to nature is really as critical and rational as he makes it out to be? I believe that we should not.

Our well-developed, scientific understanding of nature and our naturally skeptical attitude toward superstition make it difficult to appreciate the enormity of the task Henry would have faced had he decided to be more critical toward superstition. To begin with, had he not equated superstition with popular belief, then with what could he have equated it? Certainly, superstition cannot be equated with occult explanations. Occult and supernatural causes were established and important parts of Henry's world view. He readily accepted what he could not see or directly experience; so to discard some explanations as more occult than others would have had no rational justification. Moreover, if Albert's tales about snakes and Pliny's stories about the elephant and hippopotamus are not accurate, what then is to become of the rest of the information given by these much-used authorities? If a sea monster is not four acres in size, then are vipers not poisonous and does the crocodile really move both of his jaws? If the viper does not plug its ear with its tail, then do cranes not guard their flocks, sheep not flee from wolves, and geese and eagles not have a natural animosity toward one another? Since Henry had firsthand experience of only a small fraction of these and the other biological facts he presents, it would have been virtually impossible for him selectively to doubt one or two stories or descriptions without calling everything he knew into question. As a medieval scientist, there was no way for him to avoid using and accepting superstition.

To be even more specific, let me return to the example raised earlier in this chapter, of the effect of garlic on a magnet. Surely Henry had access to a garlic and a magnet. Surely he could have brought the two into contact to see what would happen. But suppose that he had, and the magnet still worked; what would negative results have proven? Any one of a number of reasons could have been given to explain "what went wrong": the garlic was not fresh, the magnet was not properly rubbed with the garlic, some other influence interfered with the action of the garlic, and so on. Since he is working with occult forces anyway, one more occult "explanation" is not out of order to save authority. Besides, if this particular fact is doubted, how much more must be doubted? Must he also doubt that garlic and magnets have other powers? If a magnet is not debilitated by garlic, is it perhaps not a good medicine, as was also commonly believed? Two hundred years after Henry's time, William Gilbert (d. 1603) was not willing to go this far. On the basis of experience he rejected the superstitious belief that garlic debilitates magnets and on the basis of experience he accepted all sorts of tales "about wonderful changes in the human body" wrought by the loadstone.[75] Clearly, more than experience is needed to cut through superstition to scientific fact.

All of this is not designed to vindicate Henry for his superstitious beliefs regarding the world of living creatures. What must be realized, however, is that faced with the complexities of the living world, the very broad nature of the problems they sought to resolve, and the limited experiential resources available to them, it is little wonder that medieval scientists took refuge in their books and traditions and in many instances ignored experience. Experience is not a good guide to the complex. It serves an important propaedeutic role in rational investigations, as even the most ardent Platonist or Augustinian of the period would have admitted. But to make more of experience than this and to attempt to base all *scientia* on experiment simply makes no sense. Except for a few simple phenomena, such as falling bodies, motion, and those associated with the mathematical sciences (perspective and astronomy), experience and experiment could play only a limited role in medieval science. Thus, myth and superstition were bound to be important. Such will become even more true as we turn to the most perfect, but also the most complex, of God's creatures, man.

CHAPTER VIII

Day Six: Humanity and the Human Sciences

And God said, "let the earth bring forth the living creature in its kind: cattle and reptiles and the beasts of the earth according to their kind." And it was made in this way. And God made the beasts of the earth according to their kind and the cattle and each reptile of the earth in its kind. And God saw that it was good. And he said, "let us make man according to our image and likeness. And let him rule over the fish of the sea and the birds of the heaven and the beasts of the earth, and over all of creation, and over each reptile that is moved upon the earth." And he created man according to his image. According to the image of God he created him. Male and female he created them. And he blessed them saying, "thrive and multiply, and fill the earth, and subdue it, and rule over the fish of the sea and the birds of the heaven and over all of the animals that are moved upon the earth." And God said, "behold, I have given to you and to all the animals of the earth and to each bird of the heaven and to all things that are moved upon the earth and in which there is the living spirit all vegetation upon the earth that bears seed and all the trees that have seed of their kind in them, for they are food for you, [and] they have [them] for eating." And it was made in this way. And God saw all the things that he had made, and they were truly good. And the evening and morning were made, day six. *Genesis 1:24–31.*

The work of distinguishing and ornamenting the elements comes to an end with the creation of man. It is to this end that all of the prior acts of creation have been striving. This work has had as its objective the establishment of an abode for humanity. Now man is created, as though "he were the end and cause of the creatures of the inferior world."[1] Even with all of his shortcomings, faults, and weaknesses, man is still God's greatest creation. He is given dominion over the rest of God's works. He is "*rex et dominus*," king and lord, of the sensible world of the elements. Even more, he is the embodiment of all that there is in the universe. Man is the microcosm that mirrors the larger macrocosm of all that exists. Like angels he has an intellect and freedom; like animals he senses; like plants he grows and reproduces; and like the heavens and earth he is composed of a mixture of the elements.[2] Even if his state is not as spiritual as that of the angels, it is man, not angels or any other creature, that is created "according to God's image and likeness."[3]

With the introduction of man into creation, Henry broached the topic that occupied his attention throughout the rest of the *Lecturae*, that is, humanity. His remaining text, the second and third chapters of Genesis, provided ample opportunity to discuss fully every aspect of the human condition, from the spiritual and religious to the demonic and superstitious, and including such diverse topics as politics, society, Church organization and discipline, the trades and professionalization, morals, and ethics. His scientific interests expressed in conjunction with the work of the sixth day are essentially twofold: first, man's creation raises questions regarding his physical being and the maintenance of the health of his being through medicine; and second, man's unique attribute, his rationality, brings into focus the entire cognitive process as viewed from the standpoint of psychology. It is to these two aspects of the human condition that this chapter is devoted.

1. The Human Body and Disease

Scholastic scientists were not equipped to deal at all exhaustively with the human body. Their arts training simply did not place much emphasis on the anatomical and physiological works that were used by the medical faculty in the training of doctors.[4] Consequently, someone such as Henry, who studied theology after his arts training, probably had very little, if any, formal training in anatomy, physiology, and the other human sciences that pertain strictly to the physical organization and functioning of the human body. Even so, he seems to have had a deep respect for these sciences. He went to some lengths to mention all of the branches of medicine in his discussion of the *arbor scientiarum*[5] and, upon occasion, praised their merits in the *Lecturae*. Indeed,

> that natural philosophy, as also that particular science that deals with natural things, that is called medicine is not immoderately praiseworthy and proper, in that it investigates the powers of plant substances and other things, and even more, endeavors to cure diseases and to preserve health through [the use of] many clever [and] contrived powers.[6]

Such benefits prompted him to give to medicine the one crucial mark of approval that he had so pointedly denied to astrology and the superstitious arts, namely, societal respectability; medicine should be permitted in and its study encouraged by the state (*res publica*).[7]

A. Physiology and Anatomy

Even without the specialized training that came with a degree in medicine, Henry was familiar with "Galenic" physiology and anatomy as they were usually understood in his day (fig. 6). He draws on elements of this physiol-

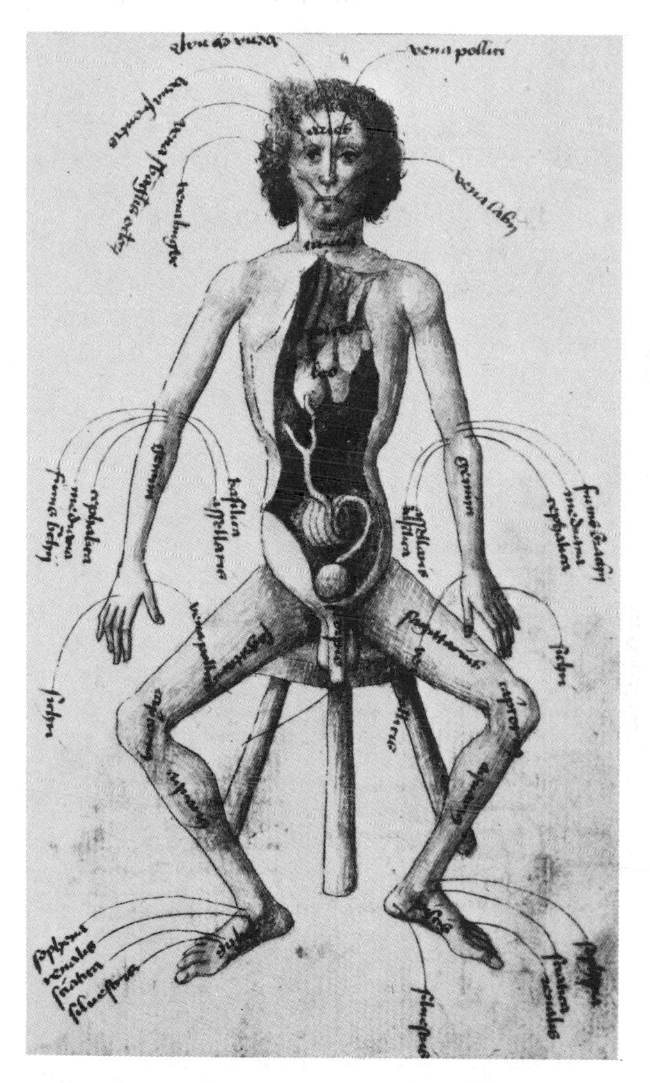

Figure 6. Human Anatomy. Depicted are the major organs–heart, lungs, stomach, intestines, bladder, and genitals–along with, among other information, the houses that govern each part. Henry would not have placed much credence in the latter.

ogy and anatomy in numerous brief discussions to resolve specific problems that arise with regard to the functioning of the human body. In these discussions the three basic groups of components that make up the human body are mentioned, either in part or in their entirety. First, in laying out the criteria for judging the perfection of animals, he notes that a perfect animal has four major organs: a heart, liver, brain, and genitals. (The fifth and final major organ that is needed to account for Galenic physiology and that is missing from this group, the stomach, is discussed in *De reductione.*)[8] Second, a brief listing of the four constituents (*complexiones*) that account for the body's overall health makes it clear that he concurs with common medical teaching and assigns four humors to the body: a sanguineous (*sanguinea*) or bloody humor, a choloric (*cholorica*) or yellow humor, a melancholic (*melancholia*) or black humor, and a phlegmatic (*phlegmatica*) humor.[9] Third, a brief discussion of the powers of the soul in conjunction with the creation of man brings in the three spirits (natural, vital, and animal, according to Henry), thus completing the basic list of ingredients needed to account for the operation of the body, as explained by Galen and his followers.[10]

The physiological system that lay behind these ingredients (a system that Henry seems to have understood only imperfectly as will be noted shortly) visualized the human body as a kind of cooking machine that takes the two basic substances coming into the body (food and air) and converts them, through the actions of the five internal organs, into the four humors and the three spirits. The three spirits are produced as food, which is first cooked in the stomach, passes into the liver and then to the heart and ultimately to the brain. At each step along the way, the food is made into an ever more subtle substance and simultaneously imbued or enlivened with spirits that are drawn in from the air and fused with the product of each cooking or "coction" (as the process of cooking was ordinarily called). The liver converts the food that it receives from the stomach into venous blood and enlivens it with the natural or nutriment-giving spirit. At the heart, venous blood is turned into arterial blood and enlivened with the vital spirit that carries the life-giving heat of the heart throughout the body. Finally, when the arterial blood reaches the brain, it is converted into a sensitive animal spirit that passes out through the hollows of the nerves and aids in sensation.[11]

The four humors arise in part from the general flow of food through the liver to the heart and brain, and in part from the actions of other organs. The melancholic humor, or black bile as it was usually called, is produced in the spleen as waste products from the stomach pass into and are absorbed by this organ. The venous blood that stems from the liver is the source of the sanguineous humor. The two remaining humors–the choloric humor or

In forming the human body, God made it perfect in every way. Legible figures are written on the features of the head. The eyebrows and bridge of the nose form the capital letter “M”; the forepart of the ear is in the shape of a lower-case “d” (the gothic “d” in Henry’s day was written “ꝺ” or “ꝺ”); the nostrils and their intervening part (*cum interstitio*) appear in the shape of a capital “E”; while the eyes make an “i” when closed and two “o’s” when open.[24] Moreover, each of the limbs is extended the same distance from the center of the body. To prove this, Henry repeats the common suggestion that a circle drawn with a compass that has one foot placed at the navel (*pes circini immobilis ponatur in umbilico et circumdetur*) would enclose exactly the ends of the outstretched hands and feet.[25] Finally, the proportions of the human body correspond precisely to another object that was inspired by God’s command, Noah’s Ark. Its proportions–300 cubits by 50 cubits by 30 cubits (Genesis 6:15)–correspond proportionally to the human body, which is six times as tall as it is wide and ten times as tall as it is from front to back.[26] With evidence such as this, there can be little doubt as to the nobility of the human body.

Such are the details that intrigue Henry when it comes to a description of the human body. His reservations about the knowability of the complexities of that body plus his desire to focus broadly on the nobility of God’s creation lead him to ignore a great deal of scientific detail and focus instead on obvious and easily verifiable characteristics. In the final analysis, he is interested in reducing the workings and disposition of the human body to their simplest components so that obvious lessons and explanations can be put forth. Even though he emphasizes again and again that the human body is complex, ultimately, when it comes to accounting for the human phenomena that interest him, the ingredients that he works with and the elements that he uses in his explanations are very simple. In no area of his consideration of the human body is this more true than in his discussion of disease.

B. Disease

Henry essentially reduces the explanation of all diseases to one simple mechanism. Given the assumption that each organ within the body has a proper primary qualitative disposition (see chap. VI, sec. 1)–that is, each has a given amount of hotness, wetness, and so on–disease can be defined as any variation from that natural disposition. When an organ’s natural qualitative balance is varied, “whether its hotness, coldness, wetness, or dryness, that part becomes ill.”[27]

There are many causes of such changes. The things that we eat, as for example rhubarb (*rheubarbarum*), which brings about a jaundiced condition (*cholera*), have an impact on the body’s natural balance.[28] Likewise, animal

yellow bile, and phlegm–were commonly thought to arise in the gall bladder and pituitary gland, respectively. Like the three spirits, the four humors also animate the body as each is associated with the qualities of one of the four elements: black bile with earth, cold, and dry; phlegm with water, wet, and cold; blood with air, hot, and wet; and yellow bile with fire, hot, and dry. Thus they provide a mechanism which accounts for the qualitative changes that take place in the body. A fever, for example, can be thought of as arising from an excess of either the sanguineous or choloric humors, both of which are hot, and in turn could be treated by the application of a cooling drug or by bloodletting.[12]

Henry’s understanding, or at least his reporting, of this physiological system is not completely in accordance with the explanation just set out. In contrast to the Galenic compartmentalization of the actions of the body in specific organs, he tends to simplify the operations of the body so that eventually almost all are reduced in one way or another to the functioning of the liver-heart-blood complex. In the case of the three spirits, for example, he reduces the functions of the vital and sensitive spirits to one “animal” (*animalis*) spirit that passes out “through the nerves and arteries as a most pure and subtle diffusible product” (*quaedam purissima atque subtilissima resolutio per nervos et arterias diffusibilis*). This reduction allows him to maintain that the body is enlivened by three spirits, while adding in addition to the natural spirit, which he accepts as outlined above, and his combined animal spirit, a third spirit, “in which there are the two souls, the intellective and sensitive” (*unum qui sunt duae animae, scilicet intellectualis et sensualis*).[13] In another passage, he further reduces all of the vital capacities of animals to the blood that arises in the liver (*jector . . . cum venis ministrat sanguinem vitamque animalis*).[14] Since venous blood is also the locus of the natural or vegetative spirit, he seems to be left with the assumption that the three Galenic spirits are all active through the blood, either venous or arterial. His added spirit for the souls is never specifically brought into the anatomical description of the body.

The reduction of all of the body’s vital capacities to the circulatory system also carries over into Henry’s description of the humors. Rather than assigning the humors to specific organs, he again seems to imagine that they are active through the blood. Thus he argues that

> the natural spirits, which are carriers of the powers or strengths of the soul, are generated from the humor of blood. And it cannot be that what is carried [in the blood] would not result in the varieties of motions of its carrier.[15]

The motions that Henry is referring to in this instance are not strictly the motions that arise from the humor of the venous blood, which would naturally be a function of that blood, but in fact motions that arise from all of the humors. Accordingly, he goes on to argue that melancholic sadness and depressions (*tristes et graves*) resulting in horrendous visions (*terribiles imaginationes*) come about from "the heaviness and coldness of melancholic blood and . . . the reception of images in the black blood" (*propter sanguinis melancholici gravitatem et frigiditatem et . . . receptionem imaginum in sanguine negro*).[16] Similarly, choloric temperaments, which express themselves in fevers and rages, derive from the blood of the heart.[17] Therefore, it seems that according to Henry, not only the spirits, but the humors too, are intimately connected with the liver-heart-blood circulatory system.[18]

That priority should be assigned to this system is made clear in Henry's discussion of the organization of the human body. Drawing on the assumption set out by Aristotle that all diversity ought to be reducible to unity and on a common assumption about the priority of organs in the human body, he concludes that all of the organs of the body are "subordinated and ordered toward one prime and principal member, which is the heart" (*est subordinatio et ordo ad membrum unum primum et principale, quod est cor*).[19] For Henry, this conclusion is not only philosophically and metaphysically true, it also conforms to physical reality. The heart *is* the origin (*principium*) of veins, arteries, nerves, other organs, blood, and the spirits. This leads to the further conclusion, also maintained by Albert, that there are not several primary organs in the body—that is, the brain as the center of the nerves, the liver as the center of the veins, and the heart as the center of the arteries.[20] Albert, of course, did not ignore the importance of the lesser organs in his discussion of the human body. Even though the liver or brain might come under the influence of the heart, he considered them on their own. However, such is not the case with Henry. The importance of the heart, which is comparable in stature to the role of the sun in the universe, seems to have dissuaded him from further investigation into the realm of anatomy.[21]

There is a second reason why Henry did not pursue an in-depth investigation of the anatomy and physiology of the human body. As was the case with animals, he regards the understanding of much of the complexity of living organisms, particularly the human organism, as beyond his comprehension. Even such basic problems as explaining how God drew the four elements together to make one body "for the most part get the better of our scientific capacities" (*longe nostram vincat scientiam*).[22] The dispositions and operations of the human body are so diverse and of so many types (*sunt valde diversae*) that there is little hope of ever presenting a complete description of them.[23] As a result, and again in close harmony with his analysis of animals, he turns to another task, that is, demonstrating how truly unique man is. And this leads him to direct his attention from the anatomy and physiology of the internal organs to the marvels of the physiognomy of the human body.

bites, such as those of scorpions and some snakes, can induce changes that may lead to death.[29] Sometimes celestial bodies have a direct influence on the human body, as when lunar rays cause a rheumatic condition (*rheumaticus*).[30] (In line with his overall rejection of astrology, Henry keeps the latter influences to a minimum; Jupiter does not cause bloody sores [*sanguinea apostemata*]; Saturn does not cause melancholy [*melancholia*]; nor do the phases of the moon cause specific diseases.)[31] Finally, natural vapors and exhalations from the earth, because of their widespread distribution, can bring about diseases of epidemic proportions.

The latter cause of disease is discussed at some length in *Contra astrologos* and brings into focus a problem that was constantly on the mind of the late medieval scientist, plague.[32] Henry attributes plague and other epidemic diseases to vapors or exhalations that arise from the bowels of the earth and travel to the middle region of the air where they are turned into a series of venomous secondary species (*species venenosae*) by the influences of the stars and the coldness of the air. These secondary species or qualities are four in number and correspond in their activities to the characteristics of the four elements: one produces melancholy as if under the influence of earth (*melancholicae quasi a dominio terrae*); another produces the choloric temperament of fire (*igneae ac cholericae*); the third gives rise to the festering fats of air (*quasi aereae putredines pingues*); and finally the fourth, to putrefying and infective dews (*rores putridi et infectivi*). This process is further refined, or complicated, by bringing intensification, remission, and the effects of the seasons to bear on the production of the exhalations and secondary qualities. Consequently, the causes of disease cannot be said to be simply four in number. Since each of the secondary qualities can be produced over a range of intensities or latitudes and since this production is further influenced by the incommensurable changes of the seasons (*incommensurabiliter*), the causes of plagues are almost infinite in number. All of which leads to the obvious conclusion that even though stellar influences are involved in the production of epidemics, the diversities that are introduced into the causal process by the actions of the elements are so complex that pestilences do not arise in any regular order in accordance with conjunctions, eclipses, or other celestial patterns.[33]

The same subtleties that are used to multiply the number of ways in which causal agents act in bringing about diseases can also be applied to the human body to explain how it reacts to disruptive qualities. Again, the overall changes involved become quite complex. In *De reductione*, Henry sets out four separate ways in which the proper balance of bodily organs can be upset: (1) Most simply, a disease can suspend one or more of the body's natural qualities, just as a magnet impedes the heaviness (the secondary quality of gravity) of iron. (2) Rather than completely overcoming one quality, a disease can cause all four qualities to be simultaneously remitted by twofold, a thousandth, or some other degree (*ex proportionali remissione omni quatuor qualitatum simul . . . ad subduplum vel submilletuplum*). (3) If all four qualities are not remitted equally, sickness can result from one or two qualities being changed and so altering the overall proportional disposition of the body or of one of its organs. Since four qualities are involved, there are six such proportions to change.[34] (4) The final change is one that relates directly to Oresme's doctrine of configurations as set out in his *De configurationibus qualitatum.*

According to the doctrine of configurations, qualities have two measurable and changeable quantities, their intensity and their extension. That is to say, qualities are not only intense, they are intense over some extended space or form, in this case, the form of the body or organ that is to become diseased.[35] This leads Henry to suggest that the

> four qualities in some member may later be equally intense as they are now when the member is well and may have the same numerical ratios as they now have but that then the animal may still be excessively sick even though it is not now sick.[36]

In other words, sickness can arise even when there is no change in the intensity (causes one and two above) or proportion (cause three) of the qualitative disposition of the organ. What then is the cause of this particular sickness?

> The cause of this is that the latitudinal intensive figuration *according to the extension* of the member which Nature determines for it can be varied in respect to one of its qualities. . . .[37]

What Henry is suggesting, then, is that it is the extended subject that changes in this case, not the intensity of the form. The organ becomes diseased because its qualitative disposition is extensively increased or decreased while its overall intensity remains the same. (Viewed geometrically, its configuration will be that of a longer or shorter rectangle of the same height.)[38] Henry does not say what such a change might represent physiologically or anatomically, or how one might diagnose such a change. He is certain, however, that explanations such as this should leave little doubt that occult powers are not needed to explain diseases since they "can be saved by the common powers of things."[39]

By reducing the actions of infectious agents to natural routes, and disease to the natural disorders of the body, Henry has set the stage for natural cures. Since disease is simply a proportional imbalance of the bodily qualities, all the physician has to do to cure the patient is prescribe medicines that will right the imbalance, which is precisely what Henry suggests they strive to do.[40] One cannot help but wonder, however, just how much efficacy he expected of this endeavor. If the above description of the ways in which the balance of the body can be changed is considered in light of an individual case—for example, even though a physician may know that a particular medicine is a laxative, he cannot know whether it will work unless he also knows how the body of his patient is "accidentally and fundamentally joined together or internally disposed" (*accidentaliter et radicaliter complexionatum aut intrinsice dispositum*)[41] —it would seem doubtful that certain knowledge of cures could ever be found. The doctrine of configuration does indeed "pass simply from the occult to the unmeasurable," as both Thorndike and Clagett have suggested.[42] If Henry's descriptions of disease were taken literally, it would seem as though the task of the doctor would be without end. By the time he gets finished accounting for disease, Henry introduces so many variables into his explanations that diagnosis and prescription seem impossible.

However, before coming down too harshly on this theory of disease, it must be recalled that Henry is primarily a theoretical and not a practical scientist. His goal is to search for causes, not cures, and in this regard his explanations are quite satisfactory. He can "save the phenomena" collectively known as disease in two important ways. First, on a general level he retains one basic mechanism that is operative in all sickness. Despite the complexities built into his theory of disease, it remains simple and all-inclusive. As with anatomy and physiology, he reduces the phenomena of health and sickness to their simplest terms, and it is on this level that cause is found. Second, his simple mechanism has built into it all of the flexibility needed to allow for the many ways in which diseases are manifest. There is sufficient latitude in his causal process to leave room for the many ways diseases occur in nature. As a theoretical physician and scientist, he has therefore met the objectives of his discipline; he has discovered cause and accounted for effect, and he has accomplished this strictly within the bounds of nature.[43] As for cures and measurement, these fell outside his domain and seem not to have been important or limiting considerations.

2. Psychology

The extraneous character of practical application and measurement to explanations of nature also applies to the case of the one human science that occupied more of the scholastic scientist's time than any other, psychology. There can be little doubt that the medieval scientist felt compelled to discuss and account for the cognitive process. Within both the context of the more strictly scientific atmosphere of the arts curriculum and the less strictly scientific atmosphere of theological studies, endless numbers of *quaestiones* were directed toward the clarification of how and what man knows. However, the actual correlation that these discussions had to physical reality remained, as did the content of the medical sciences in general, at the level of general mechanisms and commonsense explanations. Medieval psychologists were under pressure to produce explanations, not results, and to this end they devoted enormous amounts of time and energy.[44]

The investigation of the cognitive process can be carried out in a number of ways. Since the study of how we know (psychology) is closely related to the study of the nature of what we know and the knowing process (epistemology), psychological problems in the Middle Ages could be (and often were) broached in very philosophical terms, which in turn, and in line with the general orientation of scholastic philosophy, were heavily dependent on the tools of the logician. Or, since the primary agent of cognition is the soul, the study of how we know could be (and again was) approached equally as well from the standpoint of theology. This, in turn, frequently broadened the subject matter falling under the domain of psychology to include the knowing process of angels and God, topics that were commonly discussed at some length in conjunction with book one of the *Sentences*. Finally, since cognition initially begins as a physical process, psychology could be approached from the standpoint of the mechanisms and material components involved in the knowing process. This last approach had some currency in the Middle Ages, although not as much as might be expected. However, since this approach has the most direct correlation to science and nature, it is taken up in this section, as discussed by Henry in the *Lecturae* and two earlier works, his *Quaestiones super perspectivam* and *Notata de anima*.[45]

Henry bases his discussion of psychology on the epistemological assumption that everything that we know comes from outside the body to the mind (the rational soul) through the senses.[46] What he is after, therefore, as a psychologist, is an understanding of the mechanism through which the mind knows distant objects. Interest in this topic as a physical problem that takes place in and through the human body was not particularly widespread in the later Middle Ages. Although certain standard assumptions existed as to how the brain and senses function, very few of Henry's colleagues and predecessors endeavored to describe the material components of cognition with any degree of precision.[47] Galenic and Arabic discussions of cerebral psychology were ignored by all but a few commentators, such as Albert the Great, while some, notably William of Ockham, attempted to avoid the issue by definition. Ockham simply assumed that the mind "intuitively" knows distant objects through the senses, without ever explaining how.[48] Such an explanation was clearly not sufficient for Henry. In the first place, he argues, in line with Duns Scotus (d.1308) and others, that whatever intuitive capacities the human mind may have had when it was created, it lost these through the Fall.[49] In

the second place, on several occasions he engages in psychological discussions that leave no doubt that as far as he is concerned, the gap between the mind and objects must be bridged with a physical mechanism.

Bridging the gap between the mind and objects is a two-way process. In part, sensing begins with the sensitive portion of the animal spirit. Just as the action of the vegetative soul initiates the growing process and the motive portion of the animal soul brings about local motion, so too, the sensitive portion of the animal soul brings to the body the capacity to sense. As the sensitive spirits flow out through the nerves and arteries to the sense organs, they enliven these organs and render them capable of receiving information about external objects.[50] However, the sensitive spirit does not extend beyond the senses. Like most medieval scientists, Henry does not accept the Platonic notion that the senses, and particularly the eyes, have the capacity to reach out by emission and affect distant objects.[51] Therefore, there must be some action on the part of the object that completes the bridge and allows objects to be sensed.

The latter portion of the material bridge between the object and the senses is filled in by a likeness (called species or similitude) of the object. Consequently, sensing can also be said to begin when an object transmits or multiplies to the senses a likeness of itself. This likeness or species is not actually the form of the object. If it were, when it informed the matter of a sense organ, it would transform that organ into the object, which clearly does not happen. Therefore, that which is conveyed to the senses is simply a likeness of the object, and it is this likeness that completes the bridge between the senses and the object.[52] Accordingly, cognition can be described as the process by which species of objects that are multiplied to the senses act in conjunction with the sensitive spirit in sense organs to bring about sensation. Having reached this assumption, the task of the psychologist now becomes that of classifying how this simple process can account for the complex nature of the things we know.

Clarifying the psychological process begins with the five external senses (vision, hearing, smell, taste, and touch), for it is to these senses that the species of objects initially come, as experience easily confirms. Experience also makes manifest the fact that the external senses are selective. Consider the example of the oak tree raised earlier. If a tree radiates a single likeness of itself, which was usually thought to be the case, then it is clear that that likeness affects different senses in different ways: vision perceives that the tree has green leaves and a brown trunk, touch that it is hard and rough, smell that it has the peculiar odor of an oak tree, and so on. This led Aristotle and almost all subsequent commentators to agree that each sense has an aspect of the likeness of an object that it alone can and does sense, an aspect that was referred to as a "proper sensible."

> A proper sensible is that [aspect of a likeness] that is sensed by no sense other than the one [that senses it], just as color is sensed by vision and by no other [sense].[53]

However, precisely what each sense has as its proper sensible is not as obvious as might be expected.

The reason it is not always clear what specific senses properly sense is because "to move or activate" a sense can be understood in several ways. In the case of vision, for example, what the eye sees, as evidenced by the information received from the eye, is color. Color is *per se* the proper object, or proper sensible, of vision. However, if vision is reduced to its most fundamental cause, then it is clear something must precede color, since color cannot be seen without light, as was made clear above (chap. III, sec. 2). This leads Henry to the conclusion that color is *per se* but not *primo* the proper sensible of vision.[54] Furthermore, since light is itself a secondary quality, vision can also be said to be dependent on the actions of the four primary qualities: hot, cold, wet, and dry. Therefore, the process of sensation ultimately reduces to the action of the four primary qualities, which give rise to secondary qualities (light and later color in the case of vision), which in turn are multiplied to the organs of the senses and sensed as proper sensibles.[55] So depending on the question that is being asked, color, light, or the four primary qualities could be said to be the proper sensible(s) of vision.

The difficulties raised regarding the definition of proper sensibles become even more complex in the case of the sense of touch. Although it was generally agreed, after some debate, that each of the other senses has one, or possibly two, proper sensibles, touch quite obviously is not so limited. Through touch we have knowledge of the four primary qualities as well as of "certain imperfect secondary qualities such as pliability, hardness, viscosity and its opposite, gravity, levity, rarity, and density" (*quaedam secundariae imperfectae ut mollities, durities, viscositas et eius oppositum, gravitas, levitas, raritas, et densitas*).[56] Such diversity led to a number of specific problems prompting the medieval psychologist to probe more deeply into the characteristics of touch, vis-à-vis the other external senses.

Since all of the senses except touch have one or two proper sensibles, it was immediately wondered whether touch ought to be considered as a single sense.[57] The suggestion that it may not be a single sense was supported by the fact that unlike the other senses, touch has no distinct and proper organ, and by the fact that it seems to sense without an intervening medium. These difficulties led to a more general question regarding the number of external senses: Are they five in number or are there more? Finally, even with all of its imperfection, it still cannot be denied that touch has one unique attribute that may make it superior to all of the other senses; it directly perceives primary qualities. This unique attribute strongly suggests that touch may be the most reliable of the senses. Certainly it cannot be denied, as Henry concludes, that it is "the first and most natural of the senses" (*primus sensuum et naturalissimus*).[58] This conclusion did not, however, lead the medieval psychologist to place more trust in touch than in any of the other senses. Obviously, we do not receive as much information from touch as we do from the other senses. In fact, with regard to information, touch is last among the senses, after vision (the most noble of the senses), hearing, smell, and taste.[59]

As for the remaining senses—that is hearing, smell, and taste—there was little doubt that they have sound, odor, and flavor as their proper sensibles.[60] Exactly what each proper sensible is, however, had to be decided. Consequently, just as questions regarding the appropriateness of light and color to vision led to debates over what light and color are, so too, the nature of sound was explored with regard to hearing, and the nature of odors with regard to smell, and so on. As a result, there is buried within the psychological literature of the Middle Ages a wealth of information on topics relating to physics that should be, but as yet has not been, explored. Within the context of his discussion of hearing, for example, Henry explores the problem of why hard bodies, such as bells, make noise while soft bodies, such as wool and sponges, do not. This leads him into a long discussion of the nature of sound and the physical process by which it is produced. Sound is produced by "the forceful driving out of the air that is between two bodies that are brought together" (*causatur ex forti expulsione aeris existentis inter duo corpora quae concutiuntur adinvicem*).[61] Similar discussions were advanced by Henry and by most commentators on *De anima* regarding smell and taste.

The information perceived by the five external senses represents only a fraction of the total knowledge that we ultimately gain regarding objects. We know, for example, that our tree has several qualities; it is green, hard, and smells like an oak tree. Such information cannot be gained by the use of the external senses alone. Since each of the external senses knows only its proper sensible, vision cannot know that the tree is hard, touch that it is green, and so on. This fact usually led medieval psychologists to assume that in addition to the five external senses there are other senses, the so-called internal senses, that reside within the brain. These internal senses were assigned additional psychological activities that add to the complexity of the information we have regarding objects.[62]

Since the internal senses are not in direct contact with the external senses, a mechanism is needed to transfer the likenesses of objects (their species) from the external senses to the brain. This mechanism is supplied by the sensitive spirits, which are

> continually flowing from the brain all the way to the exterior [parts of the body] and thereafter reflowing by a certain motion to the common sense or imagination.[63]

To help carry out this process

> there ought to be in the hollows of the nerves that bear the sensitive spirits that go out from the brain a transparent body that is terminated at the organ of the external senses [and] appropriately illuminated for the multiplication of species.[64]

In other words, the psychological process proceeds from the external senses by means of the action of the sensitive spirits that flow out through the nerves of the body to the senses and then reflow to the brain. The nerves aid in this process since they are fitted with a transparent body capable of multiplying the species that are received and sensed in the external senses to the brain.

Once at the brain, the species can then be sensed by the internal senses that reside within the brain. There was no agreement in Henry's day as to the number of internal senses. One school of thought, which included Avicenna (d. 1037) and Albert among its supporters, conjectured that they are five in number: common sense, imagination, phantasy, an estimative power, and memory. Another school of thought, to which Henry seems to have subscribed, as did Aquinas, limited them to four in number: common sense, imagination or phantasy, a cogitative or estimative power, and memory.[65] Several psychological activities are ascribed to these senses.

(a) The common sense is the first to receive the species multiplied from the external senses to the brain. Upon receiving them, it proceeds to determine that the external senses have sensed—it was commonly believed that a sense cannot both sense and know that it has sensed—and to compose one species with another so that composite information can be derived. For example, the common sense perceives that the oak tree has green leaves, is hard, and has an ordor—qualities that none of the external senses perceived. It is also at this point in the psychological process that certain general features about objects, such as their shape, size, and so on (the common sensibles), are perceived.[66]

(b) Thereafter, the species move to the imagination or phantasy (Henry says that these powers are the same, thereby reducing the fivefold classification to a fourfold one), which stores them for brief periods of time and also derives additional, although this time nonexperiential, information, such as the image of a gold tree. It derives the latter information by drawing together the likenesses of gold and a tree and "imagining" (producing a composite image or likeness of) a gold tree.

imagination

(c) The cogitative or estimative power (again Henry makes no distinction between two powers that others distinguished) culls from the species additional bits of information that have to this point not been perceived by any other sense. For example, the estimative power of a sheep is able to perceive the hostile intent (*intentiones*) of a wolf and, as a result, flees from the impending danger. Since the estimative power is in part responsible for the sheep's flight, this power not only senses, it also motivates animals to perform certain acts, including the acts carried out *ex instinctu naturae*: building nests, spinning webs, and so on (see chap. VII, sec. 1).

(d) Finally, memory, like the imagination, stores species for future reference. However, unlike imagination, memory stores species over long periods of time. When we remember something, it is by the process of somehow dislodging the species that are stored in the memory and causing them to be remultiplied to the other internal senses. When such remultiplication takes place while we are asleep, we have dreams.[67]

The rigid compartmentalization of the activities of the internal senses was usually accompanied by an equally rigid localization of their places of activity within distinct portions of the brain. As Henry reports, members of the medical tradition distinguished three cells in the brain: anterior, middle, and posterior. The anterior cell houses either the common sense or–if in accordance with Aristotle's teaching, common sense is assigned to the heart[68] –imagination. The middle cell houses the cogitative powers, and the posterior cell houses memory.[69] However, Henry goes on to argue that he can find no evidence to assign cells to the brain; as far as he is concerned, the brain is a homogenous (*uniforme et homogeneitum*) organ. This leads him to suggest, though he seems not at all convinced of this, that if the internal senses are separable, it must be due to the fact that they reside in parts of the brain which have qualitative differences (*habet diversas partes qualitativas*).[70] This suggestion conforms to a widely held belief that the brain is cool and moist in front and dry–a proper condition for storage–in the rear.[71] However, Henry does not pursue the suggestion that there are qualitative differences, thereby leaving the problem of the localization of the internal senses unresolved.

In the final analysis and in comparison to the discussions of some of his contemporaries, Henry had very little to say in the *Notata de anima* regarding the internal senses. This is somewhat surprising, since he clearly laid a great deal of stress on the material segments of the cognitive process and spent a great deal of time in this work and in other works describing the psychological activities that precede the actions of the internal senses.[72] However, caution must be used in drawing conclusions about this topic. The concept of the internal senses is one of many concepts having no real place within the Aristotelian corpus. Aristotle only briefly discussed common sense, imagination, and phantasy, and never specifically localized these powers within the brain.[73] As a result, a commentator on the text of *De anima* who wanted to remain close to that text would have had good reason to ignore the internal senses or, as did Henry, simply mention in passing that this topic has bearing on and could be discussed within the context of Aristotle's works.[74]

Following his discussion of the internal senses in the *Notata*, Henry turns to phantasy, the one power that Aristotle did discuss at length in *De anima.*[75] Again, it is obvious from this discussion that he believes all knowledge comes from the senses. The phantasy is both a mover and something that is moved (*esse movens et motum*).[76] Like the arm that swings a stick, phantasy is "moved by sensibles" (*movetur a sensibilibus*) and moves the intellectual process that follows. As something that is moved, it is moved in accordance with that which is sensed; it faithfully represents to the intellect the information received by the senses. Thus it can truly be said, in accordance with Aristotle's definition of phantasy, that it is "a motion made from a sense according to the act of sensing" (*motus factus a sensu secundum actum . . . sentiendi*).[77] Exactly how the intellect is moved by the senses and what it knows once it is moved are topics that fall outside the domain of this section. At this point Henry directs his attention toward the first and second approaches to psychology mentioned at the beginning of this section and consequently away from psychology as a physical science.

3. The Human Machine

The image of man that emerges from this discussion of anatomy, physiology, and psychology is, in the final analysis, a very mechanical one. Man as the mirror of the macrocosm functions in much the same way as the universe at large. Cognition proceeds in a chain-like fashion comparable to, and indeed in many ways the extension of, the golden chain. The movement of spirits and humors throughout the body parallels the sweating world of meteorological exhalations and vapors. If not a clock-like machine, the human body is certainly analogous to a steam or internal-combustion engine that is driven on by vital and sensitive spirits flowing through its nerves and arteries. The medieval scientist was, it seems, bound and determined to make man one with the world of elements in which he lives.

As a machine, the human body and its functions are fairly simple and easily understood. Henry reduces both physiology and psychology to simple mechanical processes that are not at all difficult to explain. And in doing so, he grossly underestimates how truly complex the human body is. His drive to reach explanations and posit causes requires that he simplify his conceptualization of man to what we would consider the point of absurdity. Interestingly, this evaluation is not basically changed by intensification, remission and the other factors that Henry adds to the operation of the human machine to account for the many effects associated with it. Intension and remission merely add refinements to the parts of the machine; they do not change their mode of action. No matter how many degrees of intensity a disease passes through, it still acts through the same mechanism. Likewise, no matter how proper and common sensibles are defined, it is still the species acting in the senses that produces cognition. In short, no matter how complex and marvelous Henry attempts to make the human machine, he remains firmly fixed to the fundamental and common sense explanations of physiology, anatomy, and psychology that were accepted in his day.

But if the human machine is, in the final analysis, not complex, why bring complexity into its operations? There is, I would suggest, a very good reason for this. The danger of simple explanations of nature, and particularly of that aspect of nature falling under the domain of the human sciences, is that they leave room for predictions. If qualitative changes are simple and if everything that takes place within the human body is ultimately reducible to qualitative changes, then the scientist ought to be able to estimate with some certainty not only how that body functions in the present but also how it will function in the future. The psychologist ought to be able to predict what ideas will arise in the mind and the doctor forecast what impact celestial influences will have on future health. Obviously, Henry is not willing to admit such possibilities.

As a result, his objective as a scientist becomes twofold. On the one hand, he is committed to the discovery of the causes that lie behind phenomena. His explanations of humanity, and of all of nature for that matter, must be universal and complete. On the other hand, under the guise of adding mathematical exactness to the study of nature, including man and medicine, he introduces complexity into his explanations that ultimately make the realization of predictions virtually impossible. Indeed, he seems almost committed to making sure that his simple explanations do not lead to predictable results. Not only does he not foresee the possibility of theoretical knowledge leading to practical results, he seems dedicated to making sure that it does not. From our standpoint, this would, of course, undermine a major justification for doing science; however, from Henry's, it does not. As he reminds his students again and again throughout the *Lecturae*, one studies creation and God's greatest work, man, to come to better understand and appreciate the Creator, not to somehow aid in the material betterment of mankind.

CHAPTER IX

Day Seven: Rest and Reflection

> *Thereupon, the heaven and earth and all their adornment were completed. And on the seventh day God finished his work, which he had made, and on the seventh day he rested from the wholeness of his work, which he had accomplished. And he gave his blessing to the seventh day and sanctified it, because on that [day] God rested from all his work, which he had created as it had been made.* Genesis 2:1–3.

God's day of rest provides an occasion for reflection. For Henry, this meant one last opportunity to discuss the events of the prior six days and to assure his listeners that each part of the world machine (*machina mundi*) is in fact good and ordered to carry out God's commands. The distinctions between heavens, earth, matter, and form were once again set out along with a topic that had concerned him nearly two decades earlier at Paris, the golden chain. These reflections lead to a consideration of stars, cause, and effect, which, in turn, quite naturally prompted one more rebuttal of astrology.[1] Day seven also served to introduce what was to follow, that is, the extended lectures on Genesis, chapters two and three. At this point, the focus of Henry's interest shifted from science to two topics that occupied so much of his time at both Paris and Vienna: man, as a social and religious being, and society. Accordingly, at this point leave can be taken of the *Lecturae* for some reflections on their scientific content.

1. The *Lecturae* and Medieval Science: Content

The objective of this study has been to use the science of the *Lecturae* and related works as a vehicle to convey the general outlines of the late medieval *Weltanschauung*. To achieve this objective requires, of course, that the *Lecturae* and related works be a somewhat representative product of their age. In this first section of the concluding chapter, I would like to examine just how representative they are and, perhaps even more importantly, in what way they are representative of particular factions within the overall context of medieval science. Unfortunately, this examination can only begin the task that properly ought to be done. A comparative analysis of systems of thought is an exhausting and space-consuming project that lies beyond the bounds of a concluding chapter. More importantly, most of the persons to whom Henry's thought ought to be compared have yet to be the subject of comprehensive surveys such as the one just completed.[2] Given these limitations, it will still prove fruitful to set forth a few tentative conclusions, particularly to emphasize once more the representative and nonrepresentative elements of Henry's science.

Considered on its most general level, the system of science set out in the *Lecturae* can be characterized as Aristotelian. Therefore, it is typical of university science as practiced in the arts at most major European universities during the period beginning in the early thirteenth century and extending well past Henry's own lifetime. The Aristotelian elements of the *Lecturae* are unmistakable and all-pervasive: (1) With the possible exceptions of Augustine and the Bible, Aristotle is the most frequently quoted authority in the *Lecturae*. Certainly, he is the most frequently quoted authority on science. (2) The overall world view of the *Lecturae* is Aristotelian, beginning with the general homocentric arrangement of its various spheres and extending to the minute details relating to particular phenomena. (3) Most of the facts used to support this world view were originally described by Aristotle or his followers. (4) Most of the problems relating to this world view derive from Aristotle's own writings or those of his commentators. (5) The distinction of disciplines and of viable areas for investigation is shaped by the outlines of the Aristotelian corpus. (6) Even those instances in which Aristotle's ideas are corrected, as in the case of impetus, the corrections are made for essentially Aristotelian reasons. In short, there is no escaping the fact that to know and understand what the medieval scientist thought about his world, one must know and understand (not always to the letter) what Aristotle knew and thought about his world—an obvious point, perhaps, but one that cannot be overstressed.

Having said this, it must be noted immediately that to be an "Aristotelian" at this time does not mean to be a "dogmatic Aristotelian." No issue in the *Lecturae*, or in any of Henry's other scientific writings for that matter, is supported simply because Aristotle maintained it. Arguments do not run: "Aristotle said this, therefore it is true." Rather, Aristotle is brought in to add support: "this is true, as Aristotle says (*ut dicit Aristoteles*)." What comes first, then, in the order of importance is Aristotelian science, not Aristotle. Moreover, Aristotelian science represents a way of thinking that is not exclusive to the ancients, but that is common to all "right thinkers" from the time of Aristotle to Henry's own time. Aristotelian science is the "common philosophy,"[3] the common way of thinking, as taught and accepted in the schools of the ancients and in the schools of the medieval university. As such it is not a static system, but one that is subject to constant change and modification. As Aristotelians, medieval scientists worked to improve this system, not defend it. The time to retrench and protect the sanctity of Aristotelianism lies ahead in the distant future. Thinkers such as Henry are still conquering the ground that later Aristotelians will so staunchly struggle to hold.

The single most important departure in the *Lecturae* from the general outlines of Aristotelian science is toward an Augustinian interpretation of that science. Henry's Augustinian bias, if it can be called that, expresses itself in many ways: (1) The number of times that he quotes Augustine far exceeds the number of Augustinian quotations that would be found in the writings of any but the most dedicated of Augustinians of this period. Henry was obviously familiar with many of Augustine's writings and quoted them whenever possible.[4] (2) The doctrine of *seminales*, which was largely ignored if not rejected by non-Augustinian thinkers, plays an important role in Henry's thought, as in his discussion of the generation of animals. More generally, his shift away from visualizing change as a rigidly active-passive process to one of cooperation, may be seen as leaning toward an Augustinian way of thinking. (3) To raise a point not yet discussed, Henry's motivation for doing science has more in common with Augustinian thought than with the Aristotelian-Thomistic thought of the preceding century. Whereas for Aquinas the study of nature is essential for establishing a firm rational basis for approaching the supernatural, for Henry and Augustine it is more the inspirational than the rational boost one receives from nature that ultimately justifies its study. Knowledge of creation "inflames" (*inflammatur*) the soul of the scientist to admire and love the Creator. Just as one who admires and knows a picture comes to better appreciate the artist, so too, one who admires "the visible earthly machine" (*visibilis machina ista mundialis*) is led to better appreciate God.[5]

The latter orientation helps to explain why Henry devoted so much time to science in the *Lecturae* and indeed why his science, especially in his later years, takes the form it does. If science primarily provides a footing upon which to set one's logical and philosophical tools in preparation for an ascent to higher things, as is very generally the case with those who do not share Henry's Augustinian bias, then the details of that footing tend to become glossed over once their initial purpose has been realized. Once the general outlines of generation and corruption, of sublunar versus celestial, of efficient causality, and so on, have been established, the endless examples that can be brought forth to illustrate the general make-up of the world become largely superfluous. Nature serves its purpose and can be abandoned for higher activities. If, however, science and nature serve primarily inspirational roles in attaining to God, the more one indulges in nature, the more one is inspired to attain to God, which is precisely Henry's attitude toward nature in the *Lecturae*. He studies nature in all of its detail because it is God's creation. As a result, his Augustinian leanings are not only totally consistent with his science, they are, in later years at least, the primary inspiration behind this science. In the *Lecturae*, the "pursuit of science . . . has become an intensely religious activity."[6]

Placing Henry's thought in an Augustinian context narrows somewhat its proper intellectual setting. Augustinians were influential at Paris at the time he was a student there, and it is not at all surprising to find him leaning in this direction.[7] In addition to this bias, he also tends toward other intellectual currents that are more narrowly applicable to the late Middle Ages. One of these would be the tradition of intension and remission as expanded and quantified by Oresme. Another would be the use of impetus as developed by Buridan and taken over by his followers. Even though such currents are more applicable to the late Middle Ages, it is interesting to note that in Henry's mind, at least, they are assumed to be part of "the common philosophy" of the day and do not in any way set those who accept them outside the main currents of Aristotelian thought. Impetus and the doctrine of configurations are simply a part of Henry's working scientific vocabulary. He never fully explains what either concept embodies, and feels free, as in the case of Buridan's application of permanent impetus to celestial bodies, to critique them as he would any other part of Aristotle's thought, should the occasion arise.

Beyond these few specific links, what it means to be a late medieval scientist, as opposed to a thirteenth-century scientist for example, is very difficult to ascertain. The tradition of the English "*calculatores*" makes no appearance in Henry's writings, nor do any of the critical and economical approaches to knowledge that had been set out so clearly earlier in the century by William of Ockham. As we have seen, when it comes to a physical description of the process by which we know, Henry is perfectly willing to draw on species, multiplication, and all of the material precursors to thought that Ockham tried to by-pass with one simple intuitive act. Moreover, since Henry usually avoids logical and metaphysical disputes, it is impossible to determine whether he was influenced by nominalism. In short, considering content alone, there is little reason to take Henry's thought out of the general current of late medieval Augustinian-Aristotelianism and place it more narrowly into any of the supposedly more innovative schools of his day.[8]

Apart from the general similarity of Henry's science to the broad outlines of Aristotelian science and the few specific teachings that link him to Augustine, Buridan, and Oresme, there are other aspects of the content of his world view that are fairly unique. There is no clear and immediate precedent that I am aware of for his highly selective and experience-oriented discussion of cause and effect, as set out in his two specialized works on these topics and later integrated into the *Lecturae*. Nor is his extraordinary effort against astrology presaged in its scope or extent, even by the writings of Oresme. It is possible that some of his ideas on these topics stem from earlier writers, such as Roger Bacon (d. 1292), whom Henry does not quote, or William of Auvergne (d. 1249), who is quoted occasionally; however, more work needs to be done in this area.[9]

A careful study of all of Henry's early scientific and science-related works needs to be conducted using as many manuscripts as possible so that the precise content and wording of a number of central issues can be clearly established. These issues must then be traced through the preceding manuscript tradition to determine their origin. It is not until such studies are completed that a more comprehensive picture of the scope and originality of Henry's world view can be truly appreciated.

2. The *Lecturae* and Medieval Science: Method

Apart from the additions to and corrections of Aristotelian science made over the course of the fourteenth century, it is usually argued by historians that the method of doing science also changed over the same period. Reasonably speaking, then, it ought to be possible to place Henry's thought into its proper methodological context as well as into a context of ideas. In practice, the reasonableness of this suggestion is difficult to realize, since there is to date little or no agreement among these same historians as to the essence of such methodological changes.

Mathematization, probabilism, scepticism, quantification, experimentation, experiential concern, dialectical preparation, hypothetical reasoning, and so on have been advanced as possible innovative aspects of fourteenth-century thought without much agreement as to what properly ought to be concluded.[10] One reason for the lack of conclusiveness, it seems to me, is that too much time has been spent trying to determine what medieval scientists said about method and not enough on how they did science. To counter this tendency and provide an example of how method can be discussed as practiced, I have set out in this section a brief analysis of Henry's science as carried out within the pages of the *Lecturae* and related scientific works. This analysis and my reasons for not placing much stress on Henry's articulated methodology will have more meaning if we begin with a discussion of his suggestions about scientific method.

There can be little doubt about what Henry thought ought to be the method for pursuing science. As a convinced Aristotelian and one who was unmoved by Augustinian epistemology, he firmly believes that nothing can be known without experience. Knowledge derives from sense data; therefore, the scientist must begin with experiential knowledge. (In this regard, he recommends to his students that they ought to consult the beginning of the second book of Avicenna's *Canon* to discover exactly how experience should be used and how difficult it is to interpret.)[11] From experience, the scientist proceeds to a knowledge of causes. This knowledge, when it is systematized and interpreted, becomes part of the "common philosophy." Thereafter, he should advance to a knowledge of particular effects as they are administered and directed by the more general causes that comprise the "common philosophy."[12] Henry thus concurs with the general outlines of a method that was commonly set forth in his day, a method that can be traced back at least to Aristotle and, with some modifications, becomes the foundation of modern experimental science.[13]

Even though Henry suggests that this is the method that scientists ought to follow, it is not the one that he himself consistently follows. This can easily be demonstrated by taking a look at one aspect of his method, his criteria for establishing truth. If he followed an effect-cause-effect line of reasoning in pursuing science, then in theory he ought to argue with some consistency that his individual pronouncements about nature are true because they either conform to experience or are manifestations of known causes. In practice, when he is faced with actually reaching conclusions, he establishes truth in a variety of ways.

(1) *Authority.* Frequently when several plausible explanations can account for a particular phenomenon, the ultimate decision as to which is true rests on which explanation conforms to (a) the letter of Scripture, (b) what is taught publicly in the schools, (c) what is maintained by philosophers, or (d) what follows Christian teaching. Likewise, alternative explanations can be rejected because (e) they are maintained by heretics, (f) they are rejected by philosophers and theologians, and (g) they do not have public approval.[14]

(2) *Reason.* If an argument does not conform to reason, it can be rejected. Likewise, one argument can be preferred over another because it is more reasonable.[15]

(3) *Experience.* Explanations that conform to experience should be accepted. Even in this case, however, variations occur. Experience can refer to (a) saving the appearances, (b) doing so in the most apt fashion, (c) simply a probable argument, (d) an explanation that corresponds to the way things are *in esse,* or (e) one that corresponds to the way things are *in rerum natura.*[16] Experience is thus a many-sided tool that can apply to very specific phenomena (a round object remains round as it dissolves, which leads to the conclusion that the earth is round), to common experiences (water flows down), and to what can be imagined (*secundum imaginationem*) as the way things should happen.[17]

(4) *Macro-microcosm.* What happens in the larger world of the universe is

paralleled by what happens in the smaller world of the human body. Therefore, if it is known how the soul informs the body, this information can be used to reach conclusions about how intelligences are united to their orbs.[18]

(5) *Symmetry.* Since the world is God's handiwork and since God creates in an orderly fashion, it is likely that certain patterns will appear in nature. Accordingly, if all else fails, one might argue that there are twelve heavens to parallel the twelve apostles.[19]

(6) *Perfection.* Anything that is within the capacity of an inferior being must also be within the capacity of a superior being. Thus, certain things must be true of man, the most perfect of beings, because they are true of lesser animals.[20]

Clearly there is more to Henry's method than simply proceeding scientifically from experience. But if he does not follow his own articulated scientific methodology, what is the essence of the method that he does follow?

The answer to this last question lies not so much in the formal procedures that Henry follows in the *Lecturae* as it does in an attitude that he assumes in his investigation of nature. As a philosopher, scientist, and theologian, he works under the supposition that knowledge as "*scientia*" is one.[21] He assumes, without ever specifically saying so, that there is some underlying and consistent truth to the universe that may be manifest in many ways but that ultimately represents only a single truth.[22] Disciplines, as branches of the *arbor scientiarum*, can be spoken of as individual entities, as had been made very clear in the *Prologus*, but none is ever totally separable from the tree.[23] There is no reason to distinguish one branch from all others as an autonomous unit. More than this, there is no reason to limit one discipline exclusively to one method. This is not to say that some disciplines do not employ one method more than another. Certainly they do. But there is nothing to suggest that they ought to use only that method. If other disciplines have something valid to say about science, or the reverse, Henry is perfectly willing to go that route. Truth is truth no matter what its origin.[24]

The obvious and, from the standpoint of modern science, unfortunate consequence of this attitude is that the study of nature has no rules that are specifically and exclusively applicable to it alone. Whereas in theory all might agree that science is experiential and rational, in actual practice the luxury of such a limited and strict methodology was impossible to realize. This is so if for no other reason than that most medieval science begins as an historical, not an experiential, endeavor. The scholastic method and text tradition of the schools firmly grounded almost all scientific discussion in past opinion. Approaching science historically does not, of course, obviate also beginning with experience. Experience is experience, whether drawn from nature or a book. Aristotle's examples can be considered as much "the way things are" as "the way Aristotle thought that things are." But this is not the point. By beginning with texts instead of observing nature at first hand, the medieval scientist had to use the methods of the historian in his science. He had to worry about what happened when sources disagreed. He had to ponder very carefully the weight of one piece of textual evidence against every other piece of evidence. Because his experiential evidence was drawn largely from the past, it mattered whether the narrators of experience were themselves reliable. And so, scientific speculation became immediately fused with history and historical method.

Beyond this initial blurring of the bounds of scientific method, there are other reasons that explain why the medieval scientist did not adhere to a single mode of study. Medieval science was not and could not have been limited to a single approach to nature, because many things beyond nature came under its domain. Since the consideration of animals can lead to a consideration of angels and from there to demons and on to occult forces, to stars and back to the elements, it is impossible to limit method to a particular approach. As a consequence, the medieval scientist had to be scientist, metaphysician, historian, theologian, and more, all in one intellect. He had to pursue the study of nature in many ways. Methodologically, this results, of course, in utter confusion. There is no predicting at what point the historian will step forth and resolve an issue on the basis of authority, or when the metaphysician will raise a reasoned objection that may require the theologian to answer. There is, in short, no motivation or even any desire on Henry's part to limit science to what we would consider a proper scientific method.

It might be objected at this point that such ambiguity is bound to arise in an encyclopedic work that is primarily devoted to biblical commentary. However, the situation does not change very much if attention is directed exclusively to Henry's scientific works; even in these his procedures vary greatly while his articulated methodology is followed only roughly. Upon occasion he pointedly argues that experience can be deceptive or yield only approximations. We know (it is a basic precept of the Aristotelian world view) that air has only one natural movement (away from the center of the earth), even though we see it move, seemingly in a natural way, in other directions, as when it fills a vacuum.[25] We treat light rays as mathematical lines, even though they are not such in nature.[26] It would be impossible, therefore, to base science exclusively on experience. Moreover, like most of his contemporaries, Henry has frequent occasion to resort to philosophical, metaphysical, and other arguments to prove points. Key conclusions in *De habitudine* are based on the macro-microcosm and body-soul analogies.[27] Uncertain or probable arguments are employed in *Contra astrologos*, with regard to a doubt over vapor theory,[28] and in a discussion of phosphorescence in *Quaestiones super perspectivam*.[29] In the latter work, God's design in creation is brought in to help prove that the celestial fire does not give off light; if it did, it would block our vision of the heavens and frustrate the plan of creation, since we were created to walk erect so that we can see the heavens.[30] Therefore, even in his more rigorously scientific works, there is a great deal of ambiguity in the methods that are used to resolve individual problems.

There is, finally, one more general reason why Henry does not rigorously follow one scientific method, and that is because he is not seeking what a rigorous scientific method yields. The ultimate objective of his science is not a particular and highly specialized knowledge of phenomena. He is not a narrow specialist. Rather his objective is to account for specific phenomena within the large dimensions of *scientia*. Natural events have to fit into nature; nature does not conform to natural events. His science does not focus on discovery; it is preoccupied with solution. Opinions are set out and tested for the purpose of seeing which produces the most harmonious structure, which conforms best to theology, metaphysics, past opinion, and experience. To pursue one avenue of investigation and ignore all others would be an absurd activity. It would make no sense at all, as far as Henry is concerned, to crawl to the tip of one branch of the *arbor scientiarum* without keeping a firm hold on other branches for support.[31]

To summarize, it is being suggested that if viewed as actually practiced, Henry's scientific method extends well beyond his announced belief that the knowledge of cause is derived from a knowledge of effects and in turn leads to an understanding of effects. His science begins as a descriptive activity. He almost always starts by sketching out the general outlines to which his solution must ultimately conform—outlines that are called suppositions or propositions in his rigorously scholastic works. Thereafter, he brings in anomalies, the specific phenomena that are to be accounted for, usually with appropriate tentative solutions. The anomalies or individual cases are then brought into line with the larger suppositions. They are shaped and trimmed by the most appropriate methods so that ultimately they fit together as parts of some larger edifice. Finally, when the pieces are in place, the entire structure, which initially existed only in outline form, stands completed as testimony to the magnificence of the Creator. Such is the cathedral of science that is erected in the *Lecturae*.[32]

Henry does not have any naive notion that in doing this he will be able to complete the entire structure. He is well aware of the fact that many aspects of creation simply lie beyond his ken. Reaching definite conclusions about the stars is difficult, if not impossible. We may never know what the source of light was before the sun was created, how food is mixed in the stomach, or how God made animals by mixing together the four elements. Because of our fallen state, we may not even be able to know why God created the universe.[33] These and many other problems Henry leaves open to speculation. In this regard, if probabilism is a product of either the Condemnation of 1277 or the thought of Buridan or Ockham, Henry may be methodologically a product of his age. He is unwilling to commit himself to solutions in difficult areas, as was his mentor and other likely source of this reserved attitude, Augustine.[34]

What is important to note about this probabilism, however, is that it rarely extends to the crucial and guiding principles of Henry's world view. There is no probability with regard to the four elements, matter-form composition, the heavens-sublunar division, and the rest of the truly key elements of his science. About all that he is willing to admit is that these key elements can be put together in many ways. Stars might twinkle because of the atmospheric interference with their rays or because God made them that way.[35] The moon might glow during eclipses because rays pass through the earth or because the moon has some light of its own.[36] He is perfectly willing to admit that his world view, his Aristotelian world view, needs to be improved. But it would never have occurred to him to admit that his entire world view might someday have to be replaced. Why should it? He was completely satisfied that he had a proper understanding of nature firmly within his grasp.

3. The *Lecturae* and Modern Science

Henry did not, of course (from our standpoint), have a proper understanding of nature firmly within his grasp. Practically everything that he has to say about specific phenomena is by modern standards wrong and naive. In this regard, medieval science is far removed from the science of the Scientific Revolution. However, despite the niaveté of Henry's descriptions of specific phenomena, his world view is not as far removed from our modern world view as might be expected. In fact, I would go so far as to argue that if the two world views are compared on the basis of how they account for specific natural events, it is primarily details and not basic conceptualizations that separate the two.[37] Let me illustrate this point with a few specific examples. Consider for a moment medieval and modern theories regarding the origin of disease. Modern scientists describe disease in terms of bacteria or vires that infect the body and interfere with its normal functioning. The infecting agents are themselves linked to other agents, thus establishing a definite chain of events that accounts for someone's sickness. Henry's conception of the origin of disease, as discussed above (chap. VIII, sec. 1) is much the same. Plagues and other diseases are caused by stellar influences acting on the elements and giving rise to specific secondary qualities. These qualities can infect the body, thereby upsetting its normal balance and producing a particular disease. The point of this example is not that somehow Henry was on the right track when it comes to explaining disease. Obviously he was not.

The point that needs to be made is that he would not have had to change radically his own conception of disease if he were confronted by and believed the evidence used to establish modern disease theory. He basically knows that some mechanism must account for the abnormal functioning of the body and, unlike some of his contemporaries, believes that that mechanism is operative through natural channels. What he does not know, from our standpoint, is what that mechanism is. It is primarily with regard to details that his explanations of disease are wanting.

The last generalization applies not only to Henry's conception of the microcosm of the human body but also to the macrocosm of the universe. The essence of the modern world view, speaking here in terms of classical and not twentieth-century science, according to one scholar, is the belief that the world is

> . . . incapable of ordering its own movements in a rational manner, and indeed incapable of moving itself at all. The movements which it exhibits, and which the physicist investigates, are imposed upon it from without, and their regularity is due to 'laws of nature' likewise imposed from without. Instead of being an organism, the natural world is a machine: a machine in the literal and proper sense of the word, an arrangement of bodily parts designed and put together and set going for a definite purpose by an intelligent mind outside itself.[38]

This description fits Henry's world view to perfection. Because of the way his orderly and omnipotent God works in and through nature, it could not be otherwise. To imagine that God would create a lawless universe is absurd. To suggest that nature is somehow an "organism" that directs its own actions apart from God is equally as absurd. As Master Artisan, God created everything that is in the universe; he made every piece and he sets and keeps the entire structure running. God's universe must be "a machine in the literal and proper sense."[39]

Therefore, in the final analysis, and in contrast to the dichotomy of thought set out at the beginning of this book, Henry's world is mechanical. There is no other way to view a world that acts through a chain of events with the superior governing the inferior. The world is a *machina mundi*, a giant clock. Henry specifically alludes to the clock metaphor on several occasions, as has already been mentioned. God's presence may be necessary to insure that there is a clock and that it keeps running, but his presence does not, except in unusual circumstances, make the clock superfluous. Henry would be surprised to find out how we construct our clock-like universe, but he would not be surprised to find out that we believe in a *machina mundi*. Once again, it is details that stand between the medieval and modern worlds, not broad conceptualizations.

To make this point in one final way, I would contend that even on the level of hylomorphic composition, it is more details than a way of viewing the world that separates Henry's matter-form world from our world of atoms. Change in a matter-form world is just as mechanical as it is in a world of atoms. Efficient causes have every bit as much power to bring about a change as does an atom. When two things react, there is a direct head-to-head confrontation between them which ultimately results in the change that is produced. The only difference between matter-form change and atomic change is what each of the participants in change contributes. Henry's limited scientific data made it impossible for him to imagine that the physical characteristics of atoms could account for the subtleties of change; thus, he and all Aristotelians simply internalized these physical characteristics within the reactants as active and passive principles that, like atoms, direct the course of change. It is simply a matter of where the emphasis in change is placed and not that Henry's and our conceptions of change are so radically different. In this regard, one can have some sympathy with modern Thomists who have attempted to mitigate somewhat the assumed contrast between Thomism and modern science.[40]

Details are, of course, very important. In many ways the essence of the Scientific Revolution is the setting forth of specific "correct" notions, which in aggregate present a convincing case for a continued pursuit of more details. Likewise, it cannot be denied that it is Henry's consistent record of giving (what we consider to be) incorrect explanations that makes the world view previously described seem so strange and "unscientific." All of which leads one to wonder how Henry could have spent so much time pursuing details without ever realizing that his explanations were inadequate. This is really not so strange, however, if his lack of accomplishments is viewed within the context of his method.

Henry is and isn't interested in the fine points of his description of the world. He is interested in reaching what he regards to be a satisfactory explanation of particular phenomena. His world must be neat and orderly. If it is to operate with machine-like precision, the parts must fit together. However, he is not interested in the specifics relating to nature on their own. He does not pursue details in order to discover what nature is all about, but rather to solve the enigma of how the machine, which he already has firmly in view as far as general outlines are concerned, fits together. He has, in brief, none of the data gathering instincts of a modern scientist. His priorities are wrong. The whole context in which he pursues science is motivated and directed by different concerns. He does not find new answers and new solutions because he is basically not looking for them.

As strange as this might seem, I am suggesting that it is not the narrowness of scholastic science that sets it apart from modern science, but rather its universalism. To turn Henry into a modern scientist, one would have to completely reverse his priorities. The extraneous concerns and methods that fill his scientific discussions need to be stripped away. His preoccupation with larger systems and *scientia* must be replaced by a willingness to pursue details outside of their larger context. And perhaps most importantly, there needs to be added to his temperament a realization that a firm understanding of the particulars of nature can lead to tangible goods that can serve as a justification for the knowledge that produced them. Such are not the attributes of Henry's mind. However, once they become the attributes of later minds, they will produce an intellectural movement that has since gone unchecked and, indeed, may be uncheckable. For Henry's part, he found his solace and comfort in other endeavors.

Is Henry of Langenstein typical of his age? I think that he is; however, this assumption must remain for the present unproven. His own thought as expressed in the *Lecturae* needs to be more firmly grounded in the Parisian origins from which it arose, and from here the threads of influence must be traced back through intellectual movements and ages to their various origins. The one suggestion I would make for anyone intent on pursuing this endeavor is that boundaries should not be set too narrowly. Just as Galileo was anxious to be known for his tidal theory and Kepler for his insights into the harmonies of the universe, so too, medieval scientists should be known for all they wrote and did, and not simply for their occasional breaks from tradition. It is only when a scientific mind is observed in daily operation, as in the pages of the *Lecturae*, that its personality emerges. On this basis, I suspect that Henry had much company in his day.

Notes

Introduction

1. John Buridan, *Quaestiones super libris quattuor de caelo et mundo*, trans. Marshall Clagett, *The Science of Mechanics in the Middle Ages* (Madison, Wis., 1959), pp. 594–595.

2. The most obvious implication of this statement, an implication with which I am in entire agreement, is that historical analysis must begin with detailed studies of particular historical events (horizontal history) before taking up questions relating to cause (verticle history). For an interesting example of an application of this approach to intellectual history, see the works of Michel Foucault, particularly *Les mots et les choses* (Paris, 1966), and *L'Archéologie du savoir* (Paris, 1969). I have found Foucault's works very stimulating in developing my own ideas on method, although it should be noted that I make no pretense, nor do I think that it is entirely possible, to follow the linguistic approach to intellectual history that is so central to his archeological method.

3. A. Ruppert Hall in Marie (Boas) Hall, *The Scientific Renaissance, 1450–1630* (New York, 1962), p. ix.

4. Charles C. Gillispie, *The Edge of Objectivity* (Princeton, 1960), pp. 13, 16.

5. Francesco Petrarch, *De ignorantia*, trans. Hans Nachod, *The Renaissance Philosophy of Man* (Chicago, 1948), especially p. 108 ff.

6. Eugenio Garin, *Science and Civic Life in the Italian Renaissance*, trans. Peter Munz (Garden City, N.Y., 1969), pp. 2, 19, 147, 149, 153, to mention the most obvious examples. Garin, perhaps even more than most historians, should be able to avoid such pitfalls, as he has also written on the Middle Ages, e.g., *Medioevo e Rinascimento* (Bari, 1954).

7. Edwin A. Burtt, *The Metaphysical Foundations of Modern Science*, rev. ed. (Garden City, N.Y., 1954), pp. 15–24.

8. Hugh Kearney, *Science and Change, 1500–1700* (New York, 1971), pp. 23, 27. The changes that Kearney attributes to "final" cause are in point of fact discussed under "formal" cause by most Aristotelians.

9. A detailed biographical essay would be needed to prove this point. In lieu of such an essay, let me just note that what I intend to suggest in this regard is that historians of medieval science have concentrated their attention on those sciences (mathematics, physics, astronomy, and to a lesser extent the biological and medical sciences) and those problems within individual sciences (motion, gravity, and so on) that occupy center stage in the Scientific Revolution. In comparison to the number of studies devoted to these subjects, works on other and, in terms of the number of manuscript pages

written in the Middle Ages, more important or popular aspects of medieval science come in a poor second.

10. C.S. Lewis, *The Discarded Image* (Cambridge, 1964), and Brian Stock, *Myth and Science in the Twelfth Century: A Study of Bernard Silvester* (Princeton, 1972). The most comprehensive surveys of late medieval science would be either Edward Grant, *Physical Science in the Middle Ages* (New York, 1971), or, for the scholar who is interested in reading primary sources, Edward Grant, ed., *A Source Book in Medieval Science* (Cambridge, Mass., 1974).

Chapter I. The Life, Training, and Works of a Medieval Scientist

1. The facts relating to Henry's life have been discussed in some detail in a number of secondary sources, beginning with Otto Hartwig's *Henricus de Langenstein dictus de Hassia: Zwei Untersuchungen über das Leben und die Schriften Heinrichs von Langenstein* (Marburg, 1857), and extending most recently to Justin Lang's *Die Christologie bei Heinrich von Langenstein: Eine dogmenhistorische Untersuchung*, Freiburger theologische Studien 85 (Freiburg, 1966): 1–30. Lang's study contains a survey of intervening secondary literature (pp. 2–3, nn. 2, 3) and very carefully weighs the controversial evidence surrounding specific dates and facts. Since my primary concern in this study is Henry's science, I have discussed his life in these terms, omitting in the process most of the references that have bearing on specific events. The reader interested in his life would do well to consult Lang's introduction and the other relevant studies mentioned in the bibliography at the end of this book. A forthcoming edition of Henry's letters by Astrik Gabriel should add greatly to our understanding of his *curriculum vitae*.

2. J. Lang, *Christologie*, p. 19.

3. Strictly speaking, there is no medieval counterpart to modern science. Their divisions of *scientia* or *philosophia* group studies in ways that either exclude disciplines that we include in science or include disciplines that we exclude from science. Natural philosophy, for example, does not in a medieval classification of the sciences include mathematics, thus rendering it more narrow in meaning than our term science (see my article, "A Late Medieval *Arbor scientiarum*," *Speculum* 50 [1975]: 245–269). Throughout this book I have used "science" to refer generally to the study of nature. The use of this term is adopted for convenience (to avoid the more awkward phrase "the study of nature") and is not intended to ground the origins of modern science in the Middle Ages.

4. The arts program of the medieval university has been the object of numerous studies. For details concerning curriculum, see Louis Paetow, *The Arts Course at Medieval Universities with Special Reference to Grammar and Rhetoric*, University of Illinois Studies 3, no. 7 (Urbana, Ill., 1910); Hastings Rashdall, *The Universities of Europe in the Middle Ages*, new ed., F.M. Powicke and A.B. Emden (Oxford, 1936), 1: 439–471; James A. Weisheipl, "Developments in the Arts Curriculum at Oxford in the Early Fourteenth Century," *Mediaeval Studies* 28 (1966): 151–175; and Edward Grant, *Physical Science in the Middle Ages*, pp. 20–35. The trivium and quadrivium together make up the seven liberal arts. The trivium includes grammar,

rhetoric, and logic; the quadrivium, arithmetic, geometry, music, and astronomy. Lectures on these and related subjects comprised the curriculum of the arts program.

5. These requirements are found in Heinrich Denifle and Aemilio Chatelain, eds., *Chartularium Universitatis Parisiensis* (Paris, 1891), 2: 672ff. (hereafter cited as *Chart.*). Technically speaking, students received only one "degree" in the arts, the master's degree. My use of this term is merely for convenience and should not be misinterpreted as implying more.

6. Heinrich Denifle and Aemilio Chatelain, eds., *Auctarium chartularii Universitatis Parisiensis* (Paris, 1894), 1: 279, 284, 285 (hereafter cited as *Auct.*). For the typical program followed by students at Paris, see n. 9 below.

7. *Chart.*, 2: 673. At least two years had to be completed and a third year begun.

8. *Chart.*, 2: 678. The one mathematical work specified was Sacrobosco's *Sphaera*.

9. This assumption is based on a tabulation of the degree programs followed by 770 students who determined in the schools of the English-German nation between the years 1339 and 1383 (*Auct.*, 1: 25–660). Sixty-two percent of these students went on to license, taking an average time of just over nine months; fifty-two percent went on to incept, taking six additional months. If these figures are translated into the terms of the normal academic year, it means that for every student licensing three months after determination, one licensed fifteen months (one year and three months) after determination, or for every three students licensing three months after determination, one licensed twenty-seven months (two years and three months) after determination. Since some students did take as long as two years, it can be concluded that over fifty percent licensed in the same year as determination. The estimates for inception would break down in similar fashion.

10. *Chart.*, 2: 678, par. 14 [no. 15]. As was customary with regard to most resident requirements, this one was interpreted as meaning "per duos annos complete, et attingere tertium."

11. *Chart.*, 2: 673, par. 14 [nos. 6, 10], and 674, n. 3. The only restriction was that other *studia* had to have at least six masters.

12. *Chart.*, 2: 678, par. 14 [nos. 1, 2, 3, 4, 15]. The dispensations are contained in notes appended to the original statutes (*Chart.*, 2: 280). Whether or not they reflect practices in effect when Henry was a student is not certain.

13. The arrival date of 1358, given by Konrad Heilig, "Heinrich Heimbuche von Langenstein," *Lexikon für Theologie und Kirche* (1932), 4: 924, and J. Lang, *Christologie*, p. 9, is based on the assumption that Henry satisfied sequentially the resident requirements for determination and licensing. Had this been the case, his name should appear for determination in 1360 or 1361, which it does not. It should be noted that during this period some students did take two or three years to go from determination to licensing, thus suggesting that they followed a normal sequential program and did audit the prescribed texts on the trivium. I am not suggesting that this program was abandoned, only that it was deemphasized in favor of a shortened program that may have dispensed with lectures on the trivium.

14. *Chart.*, 2: 680, par. 16 [no. 2].

15. See *Auct.*, 1: 294, 298, 348, 351, 369, 375, 387, 389, 392, 394, 399, 401, 409, 421. Most of these entries pertain to sponsoring students for degrees. The English-German nation is the national subunit within the arts to which Henry, as a German, belonged.

16. *Chart.*, 2: 673, par. 4 [no. 12]. On only one occasion—November 25, 1370—was Henry granted permission to sponsor two scholars "licet non esset regens" (*Auct.*, 1: 375).

17. For manuscripts and dating, see n. 24.

18. *Quaestio de cometa*, ed. Hubert Pruckner, *Studien zu den astrologischen Schriften des Heinrich von Langenstein* (Leipzig, 1933), p. 89.

19. Léopold Delisle, *Recherches sur la librairie de Charles V* (Paris, 1907), 1: 85–119. Oresme's association with the French Court and his translations of the works of Aristotle for Charles are well known. What is less well known and to my knowledge has never been associated with the development of scholastic science during this period are the humanistic interests of some of the other translators mentioned above. John Daudin translated Petrarch's *De remediis utriusque fortunae* into French, James Bauchant, Seneca's *De remediis fortuitorum*, and Simon of Hesdin, the works of Valerius Maximus (Henry cites the latter in *Prol.* 50vb). The influence of such works and of the general intellectual milieu of the French Court on the university community is a problem deserving of more attention.

20. *Sermo*, p. 153. For the full reference to this work, see n. 43.

21. *Sermo*, p. 155. These points are made by Henry in his proof of the proposition that the order of the four faculties with regard to dignity is first theology, then law, medicine, and lastly the arts.

22. *Postilla*, 20r. For the full reference to this work, see n. 37.

23. "Ostendi aliis Parisius quando talibus vacabam in quodam tractatu de effectibus conjunctionum." *Lect.* 1: 36va. For the full reference to this work, see n. 45.

24. The former exists only in manuscript editions (ms. Paris, Bibliothèque Nationale, Latin Ms. 16401, 55r-67v); the latter two have been edited by Hubert Pruckner, *Studien*, pp. 89–138 and 139–206, respectively. Each work is dated internally; cf. *De repro.* 62r, *Q. de com.* 1 (Pruckner, p. 89), and *Contra ast.* 1. 1 (Pruckner, p. 139). (For additional editions and manuscript copies of these and the remaining works cited in this section, see the bibliography.)

25. For a summary of the content and significance of this work, see Claudia Kren, "Homocentric Astronomy in the Latin West. The *De reprobatione ecentricorum et epiciclorum* of Henry Hesse," *Isis* 59 (1968): 269–281, and "A Medieval Objection to 'Ptolemy,' " *British Journal for the History of Science* 4 (1969): 378–393.

26. Brief surveys of the content of these works can be found in Pruckner, *Studien*, pp. 23–72; Lynn Thorndike, *A History of Magic and Experimental Science* (New York, 1934), 3: 492–501; and J. Lang, *Christologie*, pp. 32–35.

27. *Contra ast.* 1. 14 (Pruckner, p. 160). *De habitudine causarum et influxu naturae communis* ms. London, British Museum, Sloane 2156, 193vb-208va.

28. *De reductione effectuum*, ms. London, British Museum, Sloane, 2156, 116vb-130va.

29. For discussions of the content of both works, see Thorndike, *History*

of Magic, 3: 474–492; Franco Alessio, "Causalita' naturale e causalita' divina nel 'De habitudine causarum' di Enrico de Langenstein," *La Filosofia della natura nel medioevo*, Atti del Terzo Congresso Internazionale di Filosofia Medioevale (Milan, 1966): 597–604; Marshall Clagett, *Nicole Oresme and the Medieval Geometry of Qualities of Motions* (Madison, Wis., 1968), pp. 114–121; and Paola Pirzio, "Le prospettive filosofiche del trattato di Enrico de Langenstein (1325–1397) 'De habitudine causarum,' " *Rivista critica di storia della filosofia* 24 (1969): 363–373.

30. The former is extant in a single manuscript: Erfurt, Wissenschaftliche Bibliothek der Stadt, Amplonianus F. 339, 73ra-108ra; the latter exists in several manuscript editions and is the only one of Henry's scientific works that was ever printed (Valencia, 1503). One short treatise on logic, which may be by Henry, also falls into the school-lecture category: *Dici de omni* (for mss., see the bibliography).

31. *De repro.* 56r. Claudia Kren, "Homocentric Astronomy in the Latin West," p. 271, suggests that the *De motibus* cited in *De repro.* could be the lost *Theorica planetarum* mentioned in a number of works on Henry (cf. Ernst Apfaltrer, *Scriptores antiquissimae ac celeberrimae Universitatis Viennensis* (Vienna, 1740), 1: 57; F.W.E. Roth, Zur Bibliographie des H. von Heimbuche de Hassia, dictus Langenstein," *Beiheft zum Zentralblatt für Bibliothekswissenschaft* 1 (1888–1889): 97; and Olaf Pedersen, "Theorica: A Study in Language and Civilization," *Classica et Mediaevalia* 22 [1961]: 161, and "The Theorica Planetarum Literature of the Middle Ages," *Classica et Mediaevalia* 23 [1962]: 225). The possible existence of such a *Theorica* was rejected by Pruckner, *Studien*, p. 6, arguing that Apfaltrer erroneously interpreted a general reference by Georg Peurbach to "theoricas planetarum et alia quaedam in astronomia" as a specific treatise. Moreover, since all subsequent references to the *Theorica* stem from Apfaltrer–or in Pedersen's case from Pruckner without note of the latter's reservations–no incontrovertible evidence for the existence of such a treatise exists. However, Pruckner did not know of a *Tractatus de sphaera*, ms., Vatican, Biblioteca Apostolica Vaticana, Vat. Lat. 9369, 41r-50r, which may be by Henry. Quite possibly, both Peurbach's reference to a *Theorica* and Henry's own reference to *De motibus* are to the *Tractatus*. If this assumption is correct, it then places the *Tractatus* as Henry's earliest extant work and provides further evidence of his initial interest in mathematical astronomy.

32. *Expositio terminorum astronomiae,* mss. London, British Museum, Harley 941, 51r–58r; Munich, Universität Bibliothek Q738, 93r–95v. The title for this work is derived from the explicit of the Harley manuscript. The attribution to Henry is derived from a heading in the Munich manuscript. The Harley edition begins with a slightly abridged version of the Munich edition and then adds a list of astronomical terms with brief definitions (fol. 52r, "Nunc ad terminorum exponendem accedamus").

33. *Q. s. per.* 14 (Erfurt 39vb, Valencia 63v).

34. *Q. s. per.*, questions 9–11 deal with vision. Question 11, in particular, goes into psychological matters not normally found in treatises on perspective (see chap. VIII, sec. 2, where psychology is discussed).

35. *Q. de com.* 1 (Pruckner, p. 89).

36. This summary of Henry's works excludes one major treatise on

medicine, *De medicinis simplicibus,* a *Dicta* on divine causality, and three *quaestiones:* "Utrum secundum naturalem philosophiam sint aliquae substantiae seperatae . . . ," "Utrum philosophica inquisitione lumen naturae attingeret speciales articulos fidei . . . ," "Utrum quodlibet corpus durum sit alteri immediate. . . . " Aside from Henry's obvious respect for medicine as a profession, I can find no firm internal evidence to attribute *De medicinis* to him. Many of the sources quoted in this work are not quoted in his other scientific treatises. The first two *quaestiones* are closely related to theological concerns and, therefore, are probably later works, while the final *quaestio* could be an early work written as an arts lecturer (for mss., see the bibliography).

37. *Postilla super Isaiam,* ms. Erfurt, Wissenschaftliche Bibliothek der Stadt, Amplonianus F. 173, 20r–95r. It is noted in the explicit (fol. 95r) that the *Postilla* were delivered at Paris, although no date is given. Since Henry had finished lecturing on the *Sentences* by 1375 (he was *baccalarius formatus* at this time [*Auct.,* 1: 478]), it is usually assumed that his required lectures on the Bible were delivered between 1372 and 1374, and his lectures on the *Sentences* between 1374 and 1375. This dating is by no means certain, since the 1375 reference could come at the end of a required three to five year residence as *formatus* (*Chart.,* 2: 700, 39), thus placing both sets of lectures three or more years earlier. However, given the fact that Henry was still active in the arts as late as 1373, it would seem likely that he simply condensed his years of lecturing to one year on the Bible (1373–1374) and one year on the *Sentences* (1374–1375).

38. The confusion regarding the authorship of the Eberbach lectures was cleared up by Damascus Trapp, "Augustinian Theology of the 14th Century," *Augustiniana* 6 (1956): 252. Trapp's evidence was overlooked by Justin Lang, who attempted to prove, following Albert Lang, that the Eberbach lectures are Henry's own lectures delivered at Paris, ca. 1374–75 (cf. J. Lang, *Christologie,* p. 60, and Albert Lang, *Die Wege der Glaubensbegründung bei den Scholastikern des 14. Jahrhunderts,* Beiträge zur Geschichte der Philosophie des Mittelalters 30, nos. 1, 2 [Munich, 1931]: 214).

39. *Quaestiones quarti Sententiarum,* Vienna, Österreichische Nationalbibliothek, CVP 4319, 145r–237v, is incomplete–the lectures on book one and part of book two are missing–and includes references to Henry of Oyta (fols. 183r-v, 187v) that are not in the earlier manuscript copy: Alençon, Bibliothèque de Ville 144, 1ra–140va. Alençon 144 is by all appearances complete; however, J. Lang, *Christologie,* p. 64, feels that the lectures contained therein are not extensive enough to be student lectures.

40. *Tractatus de discretione spirituum,* Vienna, Österreichische Nationalbibliothek, CVP 5086, 99r–108v. For brief discussions of the content and significance of this work, see Konrad Heilig, "Kritische Studien zum Schrifttum der beiden Heinriche von Hessen," *Römische Quartalschrift* 14 (1943): 125–126, nn. 3, 4; Thorndike, *History of Magic,* 3: 503–505; and François Vanderbroucke, "Henri de Langenstein," *Dictionnaire de spiritualite* (1969), 7: 218.

41. *Tractatus de horis canonicis,* ms. Munich, Bayrische Staatsbibliothek, CLM 5338, 199r–206v. Heilig, "Kritische Studien," p. 150, dates this work to the Eberbach period on the basis of astronomical data.

42. *Tractatus Venerabilis Magistri Hainrici de Hassia contra quendam Eremitam . . . nomine Theolophorum,* ed. Bernard Pez, *Thesaurus anecdotorum novissimus,* Augsbourg, 1721, 2: 507–564.

43. *Sermo de Sancta Katharina Virgine,* ed. Albert Lang, "Die Katharinenpredigt Heinrichs von Langenstein," *Divus Thomas* 26 (1948): 132–159.

44. Of interest in this regard would be *De nativitate beatae Mariae Virginis* (1390), *De conceptione beatae Mariae Virginis* (1389), *De assumptione beatae Mariae Virginis* (1385), and *De ascensione Domini* (1390), ms. Erfurt, Wissenschaftliche Bibliothek der Stadt, Amplonianus Q. 150, 128r–142r, 154r–162r, 168r–191v, 209r–233v, respectively. I have not used the sermons in this study and hope to spend more time on them in the near future.

45. *Lecturae super Genesim,* mss. Vienna, Österreichische Nationalbibliothek, CVP 3900, 146ra–395v; CVP 3901, 350 fols.; CVP 3902, 254 fols. CVP 3900, 1ra–135vb contains the *Lecturae super prologum.* Thereafter, the foliation begins again at 1ra with the *Lecturae super Genesim* and runs through fol. 243v. I have used the latter foliation in all references to *Lect.* 1. In preparing this study, I frequently checked readings in *Lect.* 1–3 with other manuscript editions, finding, in most instances, remarkable consistency. Since the objective of this book is to provide a general survey and not a critical edition, I have not included variant readings, as these would have further complicated the notes without changing my arguments.

46. "Et cum a principio Genesis usque ad capitulum quartum tangatur totius creaturae originatio et originalis constitutio, intendo, Domino concedente, circa illum passum Sacrae Scripturae aliquantulum stare in consideratione creaturarum in ordine ad Deum et Dei ad ipsas." *Lect.* 1: 10vb.

47. Both prologues derive from introductions to the Bible written by Jerome. These introductions became associated with the Vulgate as early as the ninth century and were commented upon along with the Bible (Samuel Berger, "Les préfaces jointes aux livres de la Bible dans les manuscrits de la Vulgate," *Memoires présentes pars divers savants à l'Académie des Incriptions et Belles-lettres* 1st ser. 11, pt. 2 [Paris, 1904]: 21, 33). Henry's immediate source for the organization of his *Lecturae* is Nicholas of Lyra, whom he quotes verbatim at several points (cf. *Lect.* 1: 11ra–13rb, 20ra, and Nicholas of Lyra, *Postilla super totam Bibliam* [Nuremberg, 1494], 21va–23rb).

48. A reference to 1390 appears in *Lect.* 2:267ra, "sicut . . . est homine 1390°."

49. *Lect.* 3: X2rb [240rb]; for the Latin text, see chap. IV, n. 63.

50. For example, see J. Lang's brief analysis, *Christologie,* pp. 69–73, 205–234.

51. *Lect.* 1: 159ra–va, 3: X9va [247va].

52. "Qui enim non cognoscit aut non advertit ingenium operandi artificis aut subtilitatem artificiosi operis minus miratur sapientiam factoris . . . ." *Lect.* 1: 19va–vb.

53. For surveys of this literature, see Frank Robbins, *The Hexaemeral Literature: A Study of the Greek and Latin Commentaries on Genesis* (Chicago, 1912), and E. Mangenot, "Genèse," *Dictionnaire de théologie catholique* (1915), 6: 1206–1208.

54. Of interest in this regard are the works of Mircea Eliade, particularly *The Myth of the Eternal Return,* trans. Willard R. Trask (New York, 1954),

and the recent study by Stanley Jaki, *Science and Creation: From Eternal Cycles to an Oscillating Universe* (New York, 1974).

55. For editions of these works, see Robbins, pp. 187–189; Mangenot, 1207–1208; and Friedrich Stegmüller, *Repertorium Biblicum Medii Aevi,* 7 vols. (Madrid, 1951), alphabetically listed by author.

56. Thomas Aquinas *Summa theologicae* 1a. 44–102, most of which deals with angels and man.

57. For a discussion of the development of *Sentence* commentary literature during the fourteenth century, see Palémon Glorieux, "Sentences (Commentaries sur les)," *Dictionnaire de théologie catholique* (1941), 14: 1875–1884.

58. For a discussion of Augustinian currents in the fourteenth century, see Trapp, "Augustinian Theology," and William Courtenay; "Nominalism and Late Medieval Thought: A Bibliographical Essay," *Theological Studies* 33 (1972): 730–731, and "Nominalism and Late Medieval Religion," in *The Pursuit of Holiness in Late Medieval and Renaissance Religion,* ed. Charles Trinkaus with Heiko Oberman (Leyden, 1974), pp. 51–53.

59. For a discussion of the content of these and other medieval encyclopedias, see Robert Collison, *Encyclopaedias: Their History throughout the Ages* (New York, 1964), pp. 44–81.

60. If the limits of creation are extended to include the second week, then Genesis also provides an outline for discussing the domestic and social sciences, topics usually dealt with at the end of medieval encyclopedias.

61. Vincent of Beauvais *Speculum naturale* Prol., c.1 [Strassburg, about 1481], no foliation.

62. This division is maintained despite the fact that Henry, following Nicholas of Lyra, divides God's creative acts into the work of distinction (days one through three) and the work of ornamentation (days four through six) (see chap. II, n. 13). The justification for including the fourth day with the first three seems to be that it helps retain the unity of astronomical concerns.

63. *Lect.* 1: 11va–vb; cf. Nicholas of Lyra *Postilla* 21va.

64. Cf. *Postilla* 21va ff. Nicholas does not identify the author of this exposition. It is suggested in the *Lecturae* that it is Jerome: "et hoc secundum duas expositiones . . . Hieronymo et Augustino, quarum prima . . . processit secundum duas opiniones famosas de materia prima, secunda . . . secundum opinionem tertiam." *Lect.* 1: 46vb. The first exposition is closer to Augustine than any other exposition, thus leading to the assumption that the names may be reversed, although I can find no justification for assigning the second exposition, which clearly follows Bede and Strabo (n. 65), to Jerome.

65. Cf. Nicholas of Lyra *Postilla* 23rb; Bede *Hexaem.* 1 (PL 91: 13–14), and Walafrid Strabo *Glossa ordinaria, Liber Genesis* 1 (PL 113: 67–70).

66. It is possible that Henry's fourth section is following yet another exposition, as he begins: "et consequenter per litteram secundum istam quartam expositionem." *Lect.* 1: 53vb. If so, I have been unable to identify his source.

67. The complex nature of the *Lecturae* undoubtedly explains why so many of the manuscript editions contain indices or *tabulae.* I have not

systematically investigated these indices, although they are certainly worth attention. The varied emphases and coverage of each—they are not the same—would provide an insight into the parts of and topics within the *Lecturae* later scholars found most interesting.

68. For a discussion of the origin of the sciences, see *Prol.* 56vb–57rb, and *Sermo,* pp. 144–146.

69. In setting out this threefold view of history, I strongly maintain that current scholarship on medieval historiography in its drive to evolve a simple model for the medieval schoolman's vision of the past does not do justice to this vision. If historiographical thinking does develop from the early medieval belief that past and present are not distinct, to the Renaissance belief in a distinction between the two, I would then argue that late medieval historiography, as reflected in the writings of scholastic thinkers such as Henry, must be seen as a transition period in which both beliefs are present. Accordingly, the thesis so often expressed that Renaissance historiography evidences a sharp break from its medieval counterpart (see Beryl Smalley, *Historians in the Middle Ages* [London, 1974], pp. 192–193; Peter Burke, *The Renaissance Sense of the Past* [London, 1969], pp. 1–7; and Eugenio Garin, *Science and Civic Life,* pp. 18–19) seems to me to be inadequate and misleading.

70. Phrases such as "ut modus philosophiae Peripateticae" (*Lect.* 1: 49ra), "vero secundum Peripateticos" (*Lect.* 1: 89rb), or "patet per omnes perspectivos" (*Q. s. per.* 1 [Erfurt 29ra, Valencia 47r]) are common in Henry's writings and frequently form an important part of crucial arguments. As will be discussed later (chap. IX, sec. 2), Henry is quite willing to rest his case on the weight of an opinion commonly held by the members of a discipline.

71. For examples, see *Lect.* 1: 2va, 13rb, 25va, 41vb, 47va, 49ra, 52rb; *De red. eff.* 121ra, 122vb, to mention only a few.

72. *Lect.* 1: 21ra–rb; cf. Hugh of St. Victor *De sacramentis* 1. 1. 4 (PL 176: 189).

73. *Lect.* 1: 4va, where Origen, Gregory, and Ambrose are referred to as ancients.

74. *Prol.* 48rb, where Albumasar is listed as an example of an ancient deceived by superstition.

75. That Hugh of St. Victor would under normal circumstances be considered a modern is the clear inference that can be drawn from the passage noted above, n. 72. See also *Lect.* 1: 2va, where modern philosophers and Christian theologians are spoken of together.

76. *Lect.* 1: 41vb, where the opinions of "moderni astrologi" are referred to; 49rb, where "doctores nostri" are mentioned in reference to a current debate; 70rb, where a point is made "secundum praesentem astronomicam traditionem"; or *De red. eff.* 121ra, for mention of "philosophia Peripateticorum Parisiensis."

77. It should not be assumed from the sparse showing that Henry did not rely on contemporary sources. His extensive paraphrasing of Nicholas of Lyra's *Postilla* and the similarities of many of his ideas to those of Nicole Oresme (Thorndike, *History of Magic,* 3: 481–488; and Clagett, *Nicole Oresme,* pp. 114–119) demonstrates that he was indebted to the intellectual currents of his day. However, this does not undermine the fact that he seldom quotes his contemporaries by name in crucial arguments. Opinions that do not have the experience of years are simply not weighty.

78. See Grant's discussion of this point, *Physical Science,* pp. 83–90.

79. *Sent.* 104vb, 105ra, 105vb. For a discussion of Facinus, see Trapp, "Augustinian Theology," pp. 239–242. In his study of Henry's Christology, Lang further ties his thought to Gregory of Rimini, Thomas of Strassbourg, and Giles of Rome (J. Lang, *Chistologie,* pp. 335–350).

Chapter II. *In principio*: Matter, Form, and Metaphysics

*The text of Genesis translated here and at the beginning of each of the following chapters is taken from the *Lecturae.* The translations are my own.

1. *Lect.* 1: 13ra. The theological ramifications of this issue are taken up in the *Sentences;* see *Sent.* 48ra, "Utrum sit aliqua productio in divinis qua nec essentia nec Spiritus Sanctus nec producat nec producatur?"

2. *Lect.* 2: 10vb.

3. *Lect.* 1: 162va–vb; cf. Aristotle *De caelo* 1. 8. 276a18ff.

4. Basil *Hexaem.* 2. 8 (PG 29: 48). This suggestion is raised and rejected within the context of Augustine's discussion of night and day; cf. *Lect.* 1: 22rb and Augustine *De Gen. ad litt.* 1. 15 (PL 34: 258).

5. The importance of qualifying and properly understanding the use of *de potentia Dei absoluta* in fourteenth-century writings has been convincingly demonstrated by William Courtenay, "Covenant and Causality in Pierre d'Ailly," *Speculum* 46 (1971): 97–102.

6. *Lect.* 1: 154rb.

7. *Lect.* 1: 163ra–rb.

8. Augustine *De Gen. ad litt.* 2. 1 (PL 34:263). This suggestion is assumed by Henry when he rejects special influences in favor of the actions of the four primary qualities as the means by which events in this world ought to be explained (*Lect.* 1: 37ra, and *De red. eff.* 122va–vb).

9. Henry rejects the suggestion that God acts out of any necessity (*Lect.* 1: 154ra).

10. *Lect.* 1: 21vb. This is one of three opinions given on this subject; see the discussion that follows for the other two.

11. *Lect.* 1: 161rb.

12. Henry explains and rejects Augustine's position in *Lect.* 1: 11va, 159rb; cf. Augustine *De Gen. ad litt.* 1. 2 (PL 34: 217), 4. 33 (PL 34: 318). Augustine's allegorical interpretation rests on the assumption that the account of creation given in Genesis can be compared to angelic cognition. The point that Augustine makes is that angels know a created thing either directly in God (morning knowledge) or in its being (evening knowledge) and that this distinction is what is implied in Genesis by the terms "morning" and "evening" (Augustine *De Gen. ad litt.* 4. 29–32 [PL 34: 315–317] and *Civ. Dei* 11. 7 [PL 41: 322]). The concept of simultaneous creation was well developed by members of the Alexandrian school: Philo, Clement of Alexandria, Origen, and so forth (see William Wallace's discussion of this point in Aquinas *Sum. theol.* [BF 10: 203–204]).

13. Although Henry's immediate source for the literal interpretation and division of the six days is Nicholas of Lyra (see chap. I, n. 62), he could have found similar treatments in a number of discussions of the six days, beginning generally with the Fathers of the Church; cf. Ambrose *Hexaem.* 1. 8. 28 (CSEL 32: 26–28), Hugh of St. Victor *De sacramentis* 1. 1. 24, 25 (PL 176:

202–203), Alexander of Hales *Summa* 3. 1 (Quar. 2: 319), Aquinas *Sum. theol.* la (BF 10: 2), to mention only a few of the more obvious.

14. For specific statements regarding successive creation, see *Lect.* 1: 159rb, 2: 8va–vb.

15. Among the many discussions of this general description of the medieval world, see the very readable account given by C.S. Lewis, *The Discarded Image*, pp. 92–121.

16. The various interpretations of the opening lines of Genesis are discused in *Lect.* 1: 13ra, 20ra–22ra.

17. Aristotle *Metaphysica* 4. 2. 1003b19–23.

18. For the opinion of one modern scientist regarding the method he uses, see Michael Ovendon, "Intimations of Unity," *Science and Society: Past, Present, and Future*, ed. Nicholas Steneck (Ann Arbor, Mi., 1975), pp. 363–374.

19. Aristotle *Physica* 1. 4, 5. 187a12–189a10, Ambrose *Hexaem.* 1. 1. 1–4 (CSEL 32: 1–2), Augustine *Epistola ad Nebridium* ep. 3 (PL 33: 64). For histories of atomism, see Kurd Lasswitz, *Geschichte der Atomistik vom Mittelalter bis Newton* (Leipzig, 1890), 1: 1–259, and Léopold Mabilleau, *Histoire de la philosophie atomistique* (Paris, 1895), pp. 1–396.

20. *Lect.* 1: 48va, cf. Aristotle *Metaphysica* 1. 4. 985b3–22.

21. *Lect.* 1: 48va, cf. Aristotle *Physica* 1. 4. 187a25–30.

22. *Lect.* 1: 48va, cf. Plato *Timaeus* 53C–55C.

23. The best known advocate of atomism in the fourteenth century is Nicholas of Autrecourt (see n. 58). For a discussion of atomism in the late Middle Ages, see Grant, *Source Book*, pp. 312–324.

24. See, for example, Andrew Van Melsen, *From Atomos to Atom* (New York, 1960), p. 78. Van Melsen also suggests that "the Aristotelian minima theory offered at least as many possibilities as atomism." This assumption grossly underestimates the limitations of atomism and the tremendous utility of the Aristotelian matter-form theory as tools for accounting for experience. Given the phenomena medieval scientists sought to explain and the primitive nature of their science, I do not see how atomism could have offered a viable alternative to hylomorphic composition. For a more balanced discussion of Aristotelianism and atomism, see Robert Kargon, *Atomism in England from Hariot to Newton* (Oxford, 1966), pp. 3–4.

25. Plato *Timaeus* 56C–57C, where the four basic geometric elements are used to explain various phenomena, and Lucretius *De rerum natura*, particularly bk. 4 (trans. William J. Callaghan, *Lucretius on the Nature of the Universe* [Boston, Mass., 1964], pp. 45–58).

26. A good example in this regard is Descarte's descriptions of the physical world as derived from his corpuscular theory, see particularly *Principles of Philosophy* 4. 148–182 (Adam and Tannery, 9, 2: 284–305), where magnetism is discussed.

27. See Henri Frankfort, et al., *Before Philosophy* (Baltimore, Md., 1966), pp. 23–29, for a description of primitive conceptions of causality, and H.S. Thayer, *Newton's Philosophy of Nature* (New York, 1965), pp. 46–67, for Newton's views on God's role in the universe.

28. For discussions of the origin of speculations about how order or design is brought to nature, see R.G. Collingwood, *The Idea of Nature* (New York, 1960), pp. 29–48, and Giorgio de Santillana, *The Origins of Scientific*

Thought (New York, 1961), pp. 107–128. From Henry's standpoint, the matter-form theory that he ultimately adopts is "Peripatetic," with some Christian modifications necessitated by the errors of Aristotle.

29. *Lect.* 1: 13ra.

30. See *Lect.* 1: 89ra–rb, where Henry discusses, following Augustine, the error "Platonicorum, genus putabatur quaedam idea realis . . . ." Cf. Augustine *De Gen. ad litt.* 3. 12 (PL 34: 287–288).

31. *Lect.* 1: 11va. The explanation that follows was inserted into a verbatim reading of Nicholas of Lyra's *Postillae.* One reason for the ignorance of Henry's students could easily be the newness of the University of Vienna. With formal instruction beginning only one or two years before these lectures were given, it is easy to imagine why theology students may have had a weak background in the arts.

32. *Sermo*, p. 138.

33. *Lect.* 1: 49ra. The belief that some permanent matter or hyle underlies creation found its strongest support in antiquity in the writings of Plato and was taken up by twelfth-century Platonists such as Bernard Silvester. For discussions of hyle, see Leonard Eslick, "The Material Substrate in Plato," *The Concept of Matter in Greek and Medieval Philosophy,* ed. Ernan McMullin (Notre Dame, Ind., 1965), pp. 39–54, and Brian Stock, *Myth and Science in the Twelfth Century*, pp. 97–118.

34. *Lect.* 1: 11vb.

35. *Lect.* 1: 21rb, 49vb. Henry's assessment of the views of his contemporaries is very much in agreement with the characterization of medieval matter-form theory given by James Weisheipl, "The Concept of Matter in Fourteenth Century Science," *The Concept of Matter*, pp. 151–156.

36. *Lect.* 1: 50vb, see also 52rb.

37. *Lect.* 1: 21va.

38. The doctrine of seminal reasons as set out by Augustine posits that God established in nature *in principio* invisible potentialities, called "rationes seminales," that act in almost embryo-like fashion in the production of the world. These seeds allow God to be operative constantly in nature by working from within through the seeds (see Augustine *De Gen. ad litt.* 6. 6 [PL 34: 343] and 9. 17 [PL 34: 406]). Henry's most extended discussion of seminal reasons is set out within the context of comments on Augustine's theories regarding the generation of animals (cf. *Lect.* 1: 85rb–86vb and Augustine *De Trinitate* 3. 9 [PL 42: 877–878]) and does not apply specifically to first matter. It is therefore difficult to determine exactly how much of the discussion that follows can be applied specifically to first matter. For details and bibliography regarding Augustine's doctrine of seminal reasons, see Michael McKeough, *The Meaning of the rationes seminales in St. Augustine* (Washington, D.C., 1926), especially, pp. 113–114.

39. *Lect.* 1: 21ra; cf. Hugh of St. Victor *De sacramentis* 1. 1. 6 (PL 176: 190–192).

40. *Lect.* 1: 85rb.

41. *Lect.* 1: 85va.

42. *Lect.* 1: 21va.

43. *Lect.* 1: 11vb. It should be noted that form does not necessarily account for what makes one oak tree different from another—that is, the problem of individuation. Regarding this issue, Henry seems to concur with

the common position, making matter the principle of individuation. He argues at one point that two things that have the same substantial form "possunt habere materiales subiectas diversarum specierum" (*Lect.* 1: 47vb). He qualifies this statement briefly with regard to quantity.

44. See Weisheipl's discussion of the knowability of first matter, "Matter in Fourteenth Century Science," pp. 154–156.

45. See *Lect.* 1: 57vb, 75ra.

46. *Lect.* 1: 51rb.

47. *Lect.* 1: 48va–vb. Although Henry does not conclude that this theory maintains substantial unity, there seems to be no other way to interpret it. If substance does not change and the elements do, the only way one element can become another is for its form to change while residing in one common substance.

48. *Lect.* 1: 48vb.

49. *Lect.* 1: 48vb.

50. *Lect.* 1: 48vb–49ra.

51. Aristotle *Metaphysica* 1. 3. 984a8–10.

52. For a brief discussion of the alchemist's point of view, see Mabilleau, *Philosophie atomistique,* pp. 384–396.

53. *Lect.* 1: 48vb.

54. *Lect.* 1: 16ra.

55. This debate appears in the form of letters edited by Josef Lappe, *Nicholaus von Autrecourt: Sein Leben, seine Philosophie, seine Schriften,* Beiträge zur Geschichte der Philosophie des Mittelalters 6, 2 (Munich, 1908): 14*–30*. Nicholas's ideas set out in these letters are expanded in a later work, *Exigit ordo,* trans. Leonard Kennedy et al., *The Universal Treatise of Nicholas of Autrecourt* (Milwaukee, Wis., 1971).

56. Lappe, 28*, 2–3.

57. *Exigit,* 193 (Kennedy, p. 50). See also Lappe, 28*, 9–33, and the discussion of Nicholas's position by Julius Weinberg, *Nicolaus of Autrecourt: A Study in 14th Century Thought* (Princeton, 1948), pp. 54–61.

58. For discussions of Nicholas's atomism, see J.R. O'Donnell, "The Philosophy of Nicolaus of Autrecourt and His Appraisal of Aristotle," *Mediaeval Studies* 4 (1942): 109–116, and Lasswitz, *Geschichte der Atomistik,* pp. 255–259.

59. "Et certe modus huius philosophiae si esset usitatus et longa consuetudine approbatus, ut modus philosophiae Peripateticae, forte non viderentur (sic.) in pluribus minus verisimilis illo." *Lect.* 1: 49ra.

60. *Lect.* 1: 52va.

61. *Lect.* 1: 52vb–53ra. The list of errors given by Henry includes denying that forms can exist without subjects, that the first cause is of infinite power, that nothing can be made from nothing, and so forth.

62. Augustine *De Gen. ad litt.* 1. 1 (PL 34: 217), *Civ. Dei* 12. 15 (PL 41: 363–365), *Confessiones* 12. 9 (PL 32: 829); see also Alexander of Hales *Summa* 2 (Quar. 2: 121ff.), "De angelis," Aquinas *Sum. theol.* la. 61. 1–4 (BF 9: 204–215).

63. Strabo *Glossa* Gen. 1 (PL 113: 68). Since Henry assumes the *Glossa* is by Strabo, I have used this identification throughout. That he is not the actual author of this work is noted in a number of works, such as Beryl Smalley, *The Study of the Bible in the Middle Ages* (Notre Dame, Ind., 1964), p. 40.

64. *Lect.* 2: 8ra.

65. Angels were usually described as being pure form and not united to matter (see Aquinas *Sum.theol.* la. 50. 2 [BF 9: 11–15]). Since thinking (rationality) was commonly thought to be primarily a function of form—man's capacity to think is a function of his soul—angels were thought to have this capacity in its purest form and apart from matter, hence their designation as rational beings.

66. *Lect.* 2: 8ra–9ra. See also *Lect.* 1: 90vb, where Henry notes that the blessing "be fruitful and multiply" does not apply to angels because they do not multiply.

67. *Lect.* 2: 11ra–vb; cf. Augustine *De Gen. contra Man.* 1. 4 (PL 34: 179–180).

68. For a discussion of the arrangement of the celestial orbs, see chap. IV, sec. 2.

69. *Lect.* 1: 20ra–rb; cf. Dionysius *De coelesti hierarchia* 14 (PG 3: 321).

70. *Lect.* 1: 20rb. For a more detailed discussion of this point, see Alexander of Hales *Summa* 2. 1. 1. 3 (Quar. 2: 129–130).

71. *Lect.* 1: 91vb; cf. Aquinas *Sum. theol.* la. 50. 4 (BF 9:23).

72. *Lect.* 2: 4rb.

73. Augustine *Civ. Dei* 10. 9 (PL 41: 286–287).

74. *Lect.* 2: 5va; cf. Augustine *Civ. Dei* 11. 11 (PL 41: 327–328).

75. *Lect.* 2: 16rb–va; cf. Augustine *Civ. Dei* 11. 34 (PL 41: 347–348).

Chapter III. Day One: Light, Darkness, and Perspective

1. Martin Luther, *Lectures on Genesis,* ed. Jaroslav Pelikan, *Luther's Works* (Saint Louis, Mo.,1958). 1: 6.

2. Roger Bacon, *Opus majus* 2. 1, trans. Robert Burke, *The Opus majus of Roger Bacon* (London, 1928), pp. 36–37.

3. *Lect.* 1: 55vb; cf. Augustine *De Gen. ad litt.* 1. 38, 39 (PL 34: 260–261).

4. Grosseteste's views on light are most clearly set out in *De luce* and *De motu corporali et luce,* ed. Ludwig Baur, *Die philosophischen Werke des Robert Grosseteste, Bischofs von Lincoln,* Beiträge zur Geschichte der Philosophie des Mittelalters 9 (Munich, 1912).

5. The reasons that follow for assigning a place of importance to light can be found in virtually any hexameral treatise beginning with Basil *Hexaem.* 2. 7 (PG 29: 43–47) or Augustine *De Gen. ad litt.* 1. 16, 17 (PL 34: 257–259), among his many discussions of light, and extending through such works as Alexander of Hales *Summa* 3. 2. 2. 1 (Quar. 2: 326–335) to the *Sentence* commentary literature of the thirteenth and fourteenth centuries. Following the normal pattern, Henry raises reasons for light's nobility under the passage, " and he saw that it was good" (*Lect.* 1: 13vb, 54rb–56ra).

6. *Lect.* 1: 55rb.

7. *Lect.* 1: 55rb, and 19ra, 35vb, 45rb.

8. John Pecham, *Tractatus de perspectiva,* ed. David Lindberg, Franciscan Institute Publications, Text Series no. 16 (St. Bonaventure, N.Y., 1972), p. 24.

9. Ibid.

44. Ibid. Since the sun does not change in intensity and since the stars get their light from the sun, neither do they change in intensity, a conclusion Henry had reached earlier in the *Quaestiones* for slightly different reasons (*Q. s. per.* 1 [Erfurt 29ra, Valencia 47r]).

45. See, for example, Pecham's discussion of this point, *Per. com.* 1. 18 (Lindberg, p. 95).

46. *Q. s. per.* 6 (Erfurt 33rb, Valencia 54r).

47. *Q. s. per.* 6 (Erfurt 33va–vb, Valencia 54v–55r).

48. *Q. s. per.* 6 (Erfurt 33 va, Valencia 54v).

49. *Q. s. per.* 3 (Erfurt 31rb, Valencia 50v). See Lindberg's discussion of this point, "Theory of Pinhole Images," p. 316.

50. This is my interpretation, not Henry's.

51. *De Gen. ad litt.* 1. 2–3 (PL 34: 218–219), 4. 21 (PL 34: 311); cf. Alexander of Hales *Summa* 1. 2. 1. 5 (Quar. 2: 322).

52. *Lect.* 2: 15ra.

53. Cf. Basil *Hexaem.* 2. 7 (PG 29: 43–46), Bede *Hexaem.* 1 (PL 91: 16–17), Strabo *Glossa Genesis* 1 (PL 113; 71), Alexander of Hales *Summa* 1. 2. 1. 5 (Quar. 2: 322), and Bonaventure *Sent.* 2. 13. 1. 1 (Quar. 2: 312).

54. Basil *Hexaem.* 2. 7 (PG 29: 43).

55. Augustine *Confessiones* 12. 29 (PL 32: 821). Augustine's concern over darkness is most apparent in his discussion of the Manichean heresy, *De Gen. contra Man.* 1. 8 (PL 34: 179–180).

56. *Lect.* 1: 22ra.

57. *Lect.* 1: 13va; cf. Dionysius *De divinis nominibus* 4. 4 (PG 3: 700), Aquinas *Sum. theol.* 1a.67. 4 (BF 10: 67).

58. *Lect.* 1: 22ra. Alexander of Hales *Summa* 3. 2. 1. 5. 1 (Quar. 2: 322), notes that this position is held by Lombard in the *Sentences* and Bede. Lombard does mention the cloud theory (*Sent.* 2. 13. 2 [Quar. 2: 308]); however, I can find no reference to it in Bede's *Haxaem.*, as suggested in the Quarrachi edition of Alexander's works (Bede *Hexaem.* 1 [PL 91: 17]). Bonaventure obviously takes over the cloud theory in his writings; see Bonaventure *Sent.* 2. 13. 1. 1 (Quar. 2: 313).

59. *Lect.* 1: 13vb, 22rb; cf. John Damascene *De fide orth.* 2. 7 (PG 94: 887), Basil *Hexaem.* 2. 8 (PG 29: 46), and Bonaventure's discussion of this point, *Sent.* 2. 13. 1. 2 (Quar. 2: 314–316). Augustine rejects this position in *De Gen. ad litt.* 2. 1 (PL 34: 263).

60. *Lect.* 1: 22rb.

61. *Lect.* 1: 22rb; cf. Bonaventure *Sent.* 2. 13. 1. 2 (Quar. 2: 316). Another variation of this theory is given in *Lect.* 1: 13va.

62. *Lect.* 1: 156ra. Henry notes the number of years "a principio mundi ad Christum"; however, the figures that appear in the manuscripts do not agree. All give some variation of five thousand-plus years.

63. *Lect.* 1: 22va; cf. Bonaventure *Sent.* 2. 13. 1. 2 (Quar. 2: 316).

64. *Lect.* 1: 15rb–va. These opinions are drawn from Nicholas of Lyra who in turn is drawing on Hebraic sources; see *Postillae* 24rb.

Chapter IV. Day Two: The Heavens and Astronomy

1. *Lect.* 1: 23ra; see also 14ra, 57ra–rb.
2. See chap. VI, n. 7.

3. See chap. I, n. 32. The passages that follow are quoted from the Munich manuscript.

4. "Astronomia communiter dicta caelestium corporum quantitates considerat et eorum distantias insuper ortus et occasus; velocitatem, tarditatem et caetera motus accidentia; et eventus naturales in his inferioribus, scilicet: tempestatum, tranquillitatum, sterilitatis et fertilitatis et huiusmodi." *Expositio terminorum astronomiae* 93r.

5. Ibid.; cf. Isidore *Etym.* 3. 24–25.

6. *Expositio terminorum astronomiae* 93r.

7. Ibid. 93v. For a more detailed discussion of Henry's views on judicial astronomy, see chap. VI, sec. 2.

8. Plato *Timaeus* 40A.

9. Belief in the incorruptibility of the heavens is clearly founded on statements made by Aristotle *De caelo* 1. 3, 10–12. 269b18ff., 279b4ff. It should be noted that medieval astronomers were not uncompromising in their acceptance of this belief. They were willing to admit that the heavens can change and have changed. Incorruptibility was simply a convenient and, it should be added, extremely logical and common-sense, notion that fully captures and accounts for what we see in the heavens.

10. The three positions that follow are actually drawn from Nicholas of Lyra and expanded by Henry in the *Lecturae* (cf. *Postillae* 21va and *Lect.* 1: 11va). A similar threefold division is given by Durand of Saint-Pourçain *Sent.* 2. 12. 1 (Venice, 1571, 152vb).

11. Averroes *De substantia orbis* 2, 6 (*Aristotelis opera cum Averrois commentariis* [Venice, 1562], 9: 11E).

12. Cf. *Lect.* 1: 11vb–12ra, 46vb–47ra, and Averroes *De sub. orbis* 3 (Venice, 1562, 9: 9C–D).

13. *Lect.* 1: 12ra, 47va; cf. Aquinas *Sum. theol.* 1a.66. 3 (BF 10: 38).

14. See C. S. Lewis's discussion of this issue as it applies to the literary tradition that he is considering, *Discarded Image,* p. 95.

15. Typical in this regard would be the remarks of William Donahue, "The Solid Planetary Spheres in Post-Copernican Natural Philosophy," *The Copernican Achievement,* ed., Robert S. Westman (forthcoming).

16. Ambrose *Hexaem.* 1. 6. 24 (CSEL 32: 22–23); cf. Basil *Hexaem.* 1. 11 (PG 29: 26–27) and Bede *Hexaem.* 1 (PL 91: 13–14).

17. *Lect.* 1: 47va, and William of Ockham *Super 4 libros Sententiarum* (Lyon, 1494), 4: 2. 22B; cf. Durand of Saint-Pourçain *Sent.* 2. 12. 1 (Venice, 1571, 152vb), and Thomas of Strassbourg *Commentaria in IIII libros Sententiarum* 2. 12. 1. 4 (Venice, 1564, 154vb).

18. *Lect.* 1: 12ra, 47ra–va; cf. Ockham *Sent.* 2. 22B, Thomas of Strassbourg *Sent.* 2. 12. 1. 4 (Venice, 1564, 154vb).

19. *Lect.* 1: 18rb–va.

20. *Lect.* 1: 32vb–33rb.

21. *Lect.* 1: 33vb. This makes stars "non . . . continua sed contigua" with their orbs.

22. *Lect.* 1: 33rb.

23. *Lect.* 1: 14rb.

24. *Lect.* 1: 33rb; these characteristics are made manifest in the course of a discussion on whether the stars and orbs are the same.

25. *Lect.* 1: 16rb.

26. *Lect.* 1: 33rb.

27. ". . . sicut corpus Christi transivit per ianuas clausas et transivit caelos sine ruptura, ita Deus potuit stellas inferius factas sine tali inconvenienti in supremo loco firmamenti collocare." *Lect.* 1: 33va.

28. *Lect.* 1: 36rb; cf. Aristotle *Meteor.* 1. 8. 345a25–345b1.

29. *Lect.* 1: 35va–vb.

30. *Lect.* 1: 35va–36ra; cf. Bacon *Opus majus* 4. 4. 1 (Burke 1: 149–50).

31. *Lect.* 1: 37vb–38ra, and Pecham *Per. com.* 2. 56 (Lindberg, pp. 208–211).

32. Augustine *De Gen. ad litt.* 2. 18 (PL 34: 279–280).

33. *Lect.* 1: 38ra. These doubts about the number of stars are set out in response to the question: "Utrum constet evidenter quod sint solum septem stellae erraticae?" In arguing that there may not be, Henry brings in a number of extraneous issues, including seven theories regarding the nature of comets (see chap. V, n. 59).

34. *Lect.* 1: 13ra.

35. *Lect.* 1: 19ra; cf. Aristotle *De caelo* 2. 12. 292a4–10, where the eclipse of mars by the moon is discussed.

36. *Lect.* 1: 19ra; cf. Aristotle *De caelo* 2. 10. 291a30–291b10, for Aristotle's discussion of the ordering of the planets.

37. Augustine *Civ. Dei* 21. 8 (PL 41: 720), who in turn draws on Marcus Varro, *On the Race of the Roman people.*

38. *Lect.* 1: 36ra–rb. The reason for the partial nature of the eclipse would be due either to the relative size of the two planets or to the weakening of the shadow of the eclipse over distance.

39. See John Dreyer's discussion of this point, *A History of Astronomy from Thales to Kepler,* 2nd ed. (New York, 1953), pp. 168–169.

40. *Lect.* 1: 17vb, 158rb. Henry's discussion of this problem is based in part on a similar discussion presented by Geber (Jabir ibn Aflaḥ) in *De astronomia libri IX* 7 (Nuremberg, 1534, pp. 103–106). His use of this source is discussed below, n. 66.

41. *Lect.* 1: 19ra, 158rb. If the order of the orbs is not known, predicting their influences is at best on doubtful grounds.

42. For discussions of the history of diurnal rotation, see Pierre Duhem, *Le Système du monde* (Paris, 1910), 3: 43–162; Grant McColley, "The Theory of the Diurnal Rotation of the Earth," *Isis* 26 (1936–37): 392–402; Clagett, *Science of Mechanics,* pp. 583–589; and Grant, *Source Book,* pp. 494–516.

43. See Clagett, *Science of Mechanics,* pp. 594 and 600, for John Buridan's and Nicole Oresme's queries about the earth's immobility.

44. Augustine *De Gen. ad litt.* 2. 10 (PL 34: 271–272). Relegating what would normally be problems in Aristotelian science to the context of Augustine's work is a prime example of the "Augustinian" nature of Henry's science.

45. *Lect.* 1: 24va. The observations over long periods of time are needed to detect the precession of the equinoxes of about 1° per century. For additional information on this subject, see Dreyer, *History of Astronomy,* pp. 202–204.

46. See Oresme *Le Livre du ciel et du monde* 2. 25 (trans. Clagett, *Science of Mechanics,* p. 604).

47. *Lect.* 1: 25ra; the passages brought in to support this point include Job 37:18, Jos. 10:12–13, and 2Pet. 3:10.

48. See n. 32.
49. Celestial movers are discussed below, chap. VI, sec. 1.
50. Plato *Timaeus* 33B, 34A (Cornford, pp. 54–55).
51. Dreyer, *History of Astronomy,* p. 89.
52. For a brief but informative discussion of the problems of astronomy, see Thomas Kuhn, *The Copernican Revolution* (New York, 1957), pp. 45–77.
53. A complete description of Ptolemaic astronomy is given by Francis Benjamin and G.J. Toomer in *Campanus of Novara and Medieval Planetary Theory: Theorica planetarum* (Madison, Wis., 1971), pp. 39–56.
54. *Lect.* 1: 17va, 18vb, 23vb, 24ra, 76ra, and so forth.
55. Kren, "A Medieval Objection to 'Ptolemy'," pp. 378–393.
56. "Et apparuit Ptolomeo quod omnes apparentiae diversitatum in motibus planetarum salvari possent imaginando cuiuslibet eorum orbem totalem esse divisum in tres orbes. . . . Et secundum istam viam, si res consonet imaginationi universi, orbes septem planetarum sunt 29 vel 30." *Lect.* 1: 23vb. The section deleted goes into a brief description of the orbs and the distinction between eccentrics and epicycles.
57. *Lect.* 3: X5rb [243rb]. At one point in the *Lecturae* Henry assumes that Ptolemaic astronomy uses eccentric circles for both the sun and moon (*Lect.* 1: 17vb). Given this assumption, which is basically correct, it is hard to imagine how he can argue with any consistency that the sun has three circles. Campanus, for example, uses only a single eccentric circle for the sun (Benjamin and Toomer, *Campanus,* p. 143). Assuming five circles for the moon does agree with Campanus; however, the latter uses only four circles for Mercury and three each for the remaining planets (Benjamin and Toomer, *Campanus,* pp. 161, 213, 299).
58. See Kren's discussion of the weaknesses of Henry's criticism in "A Medieval Objection to 'Ptolemy'," pp. 379–382, and subsequent objections raised in *Lect.* 1: 76ra–rb, 3: X2rb–va [240rb–va], X6vb–X7ra [244vb–245ra]. Henry also notes at one point that there are errors in Ptolemy's data (*Lect.* 1: 42rb).
59. *Lect.* 1: 24ra.
60. *Contra ast.* 1. 8 (Pruckner, p. 151).
61. The controversy over the reality of eccentrics and epicycles is discussed by Pierre Duhem, *Le Système du monde,* 4: 142–151; and in Grant, *Source Book,* pp. 516–529.
62. Aristotle *Metaphysica* 12. 8. 1074a10–14.
63. "Si querit hoc a me, remitto eum ad tractatum quendam contra eccentricos et epicyclos, incipientem: 'Cum inferiorum cognitio, etc.,' quem Parisius feci sunt plusquam 28 anni, ostendens rationibus multis et considerationibus huius cedendi astronomia solum ad praedictum finem fictam esse et corrospondentiam in re non habere; ostendens etiam ibidem in genere circa finem illius tractatus per quem et qualis orbium motum possent realiter apparentiae in motibus planetarum salvari." *Lect.* 3: X2rb [240rb].
64. For a description and explanation of the constructive portion of *De reprobatione,* see Claudia Kren, "Homocentric Astronomy in the Latin West," pp. 269–281.
65. "Adhuc est alia astronomia quae per paucissimos orbes vult salvari omnes diversitates quae apparent in motibus et dispositionibus seu apparentiis planetarum. Et illa innititur cuidam extraneae speciei motus circularis quo sphaera stellarum fixarum quibusdam visa est moveri taliter: quod nullus eius

punctus quiescat nisi centrum, et duo opposita puncta describant duos parvos circulos perfectos et eodem modo epicycle flectuntur in latitudinem secundum Gebrum; de ista astronomia tactum est alias aliquid in tractatu de epicyclis et eccentricis." *Lect.* 1: 23vb–24ra.

66. Henry's source would seem to be Geber's *De astronomia liber IX.* However, after comparing sections of *De reprobatione* with the printed edition of *De astronomia* (Nuremberg, 1534), I am not entirely convinced that the latter accurately represents Geber's works as they would have been available to Henry. Accordingly, more work needs to be done to trace the precise origins of the construction portions of *De reprobatione.* Claudia Kren is presently at work on this problem as it relates to the astronomical tradition at the University of Vienna.

67. *Lect.* 1: 24ra.

Chapter V. Day Three: The Elements, Cosmography, and Meteorology

1. Cosmographical treatises, from antiquity on, usually included some consideration of the heavens. Properly speaking, then, cosmography does not pertain strictly to the elements, hence my use of the term "sublunar" in the title of this section. It should also be noted that there is no way sharply to distinguish cosmography from geography and a related discipline, cartography, as is well illustrated by their overlapping in Manuel Santarem's monumental study, *Essai sur l'histoire de la cosmographie et de la cartographie pendant le Moyen-Age, et sur les progrès de la géographie après les grandes découvertes du XVe siècle,* 3 vols. (Paris, 1849–52). Since Henry's discussion of the disposition of the elements focuses mostly on what can properly be called cosmography, I have used this term throughout this chapter, even though some of what is discussed below is clearly geographical in nature. For discussions of medieval cosmography, see, in addition to Santarem's work, Charles Beazley, *The Dawn of Modern Geography: A History of Exploration and Geographical Science,* 3 vols. (London, 1897–1906); and George Kimble, *Geography in the Middle Ages* (New York, 1968).

2. Cassiodorus *Institutiones* 1. 25. 1 (trans. Leslie Jones, *An Introduction to Divine and Human Readings* [New York, 1969], p. 125).

3. The most important contemporary work on cosmography is without a doubt Pierre d'Ailly's *Tractatus de imagine mundi.* For a discussion of this work, see Kimble, *Geography,* pp. 208–212.

4. See Albert the Great *De natura locorum* 1. 1. 1 (B. 9: 529), where he notes that although the ancients, including Plato and Aristotle, had an interest in cosmography, "their works (*libri*) have not come to us intact but only in parts." Clearly, he could find no obvious precedent within the Aristotelian *corpus* for devoting an entire work to cosmography and related topics. Even when Aristotle did discuss the sublunar realm in some detail, as in *De caelo,* medieval schoolmen did not always follow his lead. As one scholar has noted, medieval commentaries on *De caelo* often demonstrate little or no interest in the elements (Claudia Kren, "The *Questiones super de celo* of Nicole Oresme," Ph.D. dissertation, The University of Wisconsin, Madison, 1965, pp. xii–xvi).

5. Such would be the approach in the initial questions advanced by Henry's contemporary, Nicole Oresme, in *Le Livre du ciel et du monde* 1.

2–3 (ed. Albert Menut and Alexander Denomy [Madison, Wis., 1968], pp. 56–59).

6. Background on the history of the four element theory can be found in many histories of chemistry and alchemy; see in particular Marcelin Berthelot, *La Chimie au moyen âge* (Paris, 1893), or James Partington, *A History of Chemistry*, vol. 1 (London, 1961).

7. *Lect.* 1: 15ra, 28rb, 53va. The order in which each set of two qualities is given is the order in which each element has its qualities; that is, earth is first (*primo*) dry and second (*secundario*) cold, and so forth.

8. *De hab. caus.* 195ra–vb.

9. " . . . haec apud philosophos fatua quaestio est, sicut qua quaereretur quare haec arbor, demonstrato piro, fert pira et non poma." *Lect.* 1: 27rb.

10. Corporeality, as a property of elements, is a common nature, not a proper nature. That is to say, the form of any element makes that element into a three-dimensional body, thus rendering corporeality a common characteristic of the elements. Proper natures are specific to specific elements; only earth is cold and dry. See Henry's brief discussion of common natures, *De hab. caus.* 196rb–va.

11. Action at a distance remained a problem for scientists well into the Scientific Revolution. For a discussion of the debates that took place between Cartesians and Newtonians over the "cause" of gravity, see Alexandre Koyré, *Newtonian Studies* (Cambridge, Mass., 1965), pp. 115–184, and William Wallace's brief summary of the issues involved in *Causality and Scientific Explanation* (Ann Arbor, Mi., 1972), 1: 205–210.

12. This well-known principle is set out by Aristotle *Physica* 7. 1. 241b24–25.

13. The uncertainty over this issue was finally put to rest by Galileo, who was the first to establish firmly and authoritatively that the distance a falling body traverses is proportional to the time squared ($s = 1/2at^2$). Several writers prior to Galileo, such as Domingo de Soto, reached the same conclusion; however, none was able to resolve the issue to the satisfaction of all concerned. For additional information relating to free fall, see Marshall Clagett, *Science of Mechanics*, pp. 541–582, and William Wallace, "The Enigma of Domingo de Soto: *Uniformiter difformis* and Falling Bodies in Late Medieval Physics," *Isis* 59 (1968): 384–401.

14. The pioneering, and still in many ways most complete, study of the origin of the medieval impetus theory is Anneliese Maier's *Zwei Grundprobleme*, pp. 113–315. Pierre Duhem, *Études sur Léonard de Vinci* (Paris, 1909), 2: 193–200, had earlier uncovered a great deal of literature relevant to this topic; however, his interpretations are not always accurate. Subsequent material on impetus can be found in Clagett, *Science of Mechanics*, particularly pp. 505–540, and James Weisheipl, *The Development of Physical Theory in the Middle Ages* (New York, 1959; reprint. ed. Ann Arbor, 1971). The major problem concerning the origin of the impetus theory is the fact that it was put forth prior to the fourteenth century in works that were apparently unknown in the fourteenth century. Thus, the discovery of impetus by Franciscus de Marchia, John Buridan, and others seems to be original with these writers, even though they were not the first to use the impetus theory.

15. Aristotle *Physica* 4. 8. 215a14–17, 8. 10. 266b27–267a20. The first

of the two theories is generally referred to as the "antiperistasis theory" and has its origin in Plato's *Timaeus* 79E–80C. These explanations are required if it is assumed that a moving body does not continue to move unless acted upon by an outside force; that is to say, if the concept of inertia is unknown.

16. See, for example, Buridan's critique of the Aristotelian position in *Quaestiones super octo physicorum libros Aristotelis* 8. 12. (trans. Marshall Clagett, *Science of Mechanics,* pp. 532–538).

17. Clagett, *Science of Mechanics,* p. 524. It should be noted that impetus is not equivalent to inertia. Impetus is an active principle that keeps a body moving when it would not naturally do so. According to the inertia theory, a moving body continues to move because that is its natural condition, not because some quality within it keeps it moving.

18. Buridan suggests that impetus is a permanent quality that is weakened by outside factors, such as the resistance of the medium through which a moving object passes. Most of his followers assume that impetus is a naturally self-wasting quality (it dissipates like heat from a hot iron) and not permanent (Clagett, *Science of Mechanics,* pp. 524–525).

19. In point of fact, it may be a problem associated with the Eucharist, "Utrum in sacramentis sit aliqua virtus supernaturalis insistens sive eis formaliter inhaerens," that first led the assumed rediscoverer of the impetus theory in the fourteenth century, Franciscus de Marchia, to develop this theory (Maier, *Zwei Grundprobleme,* pp. 110–162). Henry applies impetus to the movement of clouds (*Lect.* 1: 84vb) and upon occasion assumes it as a working part of his scientific vocabulary (*Lect.* 1: 72ra; *De red. eff.* 124rb–va).

20. Clagett, *Science of Mechanics,* p. 551.

21. " . . . superflue ponerentur aliquae speciales qualitates motivae in elementis, sicut gravitas et levitas, quia . . . forma terrae . . . esset causativa motus simpliciter deorsum. . . . " *De hab. caus.* 196va. For a discussion of proper natures, see n. 10.

22. " . . . non videtur saltem esse magis ei naturalis quam motus naturalis quae dependet causaliter a qualitate quam vocant impetus." *De hab. caus.* 196vb. For a discussion of common natures, see n. 10.

23. Henry also rejects Buridan's suggestion that impetus can be used to account for celestial motions; see chap. VI, sec. 1.

24. *Lect.* 1: 14va, 26vb.

25. *Lect.* 1: 14vb, 27rb. Presumably, the last explanation is predicated upon the assumption that a mixture of light and heavy elements would be lighter than pure earth. For mention of the earth's rigidity, see Augustine *De Gen. ad litt.* 2. 11 (PL 34: 273).

26. *Lect.* 1: 14va; cf. Plato *Phaedo* 11C–13B, Aquinas *Sum. theol.* 1a. 69. 1 (BF 10: 98).

27. *Lect.* 1: 27ra; cf. Aristotle *Meteor.* 1. 14. 352a16ff., 2. 2. 354b1ff.; Ambrose *Hexaem.* 3. 4. 17–19 (CSEL 32: 70–72). Ambrose assumes that since the sun is created on the fourth day, the drying action of the third day must be due to either an act of God or the natural dryness that is in earth.

28. *Lect.* 1: 14va, 26vb; cf. Robert the Englishman's commentary on Sacrobosco, *De sphaera,* trans. Lynn Thorndike, *The Sphaera of Sacrobosco and Its Commentators* (Chicago, 1949), p. 205, for mention of divine and celestial action in causing dry land.

29. For a discussion of a few of the many opinions regarding the origins of the Deluge, see Kimble, *Geography*, pp. 154–155.

30. *Lect.* 1: 27ra. The mention of these three examples is one of the few instances that Henry cites local place names rather than using the common examples drawn from the ancients.

31. *Lect.* 1: 63vb. The example of the canal connecting the Red and Mediterranean Seas is not mentioned by Henry, but can be found in many commonly available sources, as for example Aristotle *Meteor.* 1. 14. 352b20–34, Basil *Hexaem.* 4 (PG 29: 83), Ambrose *Hexaem.* 3. 2. 11 (CSEL 32: 67).

32. *Lect.* 1: 61rb–va. Henry draws support for the overall rotundity of the earth from Sacrobosco *De sphaera* 1 (Thorndike, *The Sphaera of Sacrobosco and Its Commentators* [Chicago, 1949], pp. 121–122), although many other sources could have been used. The antipodes were thought to be lands that lie opposite (or below) our own hemisphere. Whether or not such lands exist and whether or not they are inhabited if they do exist, are problems that provoked a great deal of speculation. Isidore, for example, suggests not only that there are such lands but that they are inhabited (*Etym.* 14. 5. 17.). In contrast, Augustine completely rejects the notion that there should be other habitable lands (*Civ. Dei* 16. 7–9 [PL 41: 485–487]). Between these two opposing views, many intervening positions can be found; see, for example, Albert the Great's discussion of habitable lands in *De natura locorum* 1. 6–8 (B. 9: 538–546).

33. *Lect.* 1: 61vb.

34. *Lect.* 1: 62ra. This proportion is based on Ptolemy's estimate of 34 (sic.) to 1 as the ratio of the radius of the moon's orbit to the radius of the earth (Ptolemy *Almagestum* 5. 13 [trans. R. C. Taliaferro, *Great Books of the Western World,* Chicago, 1952, 16: 171], where 39 and not 34 to 1 is given). If it is then assumed that the surfaces of two spheres are proportional to their radii cubed, the ratio of 39,000 to 1 is derived for the approximate relationship between the surface area of the earth and the surface area of the sublunar region (the sphere enclosed by the orbit of the moon). Henry then derives from this "quatuor numeris continue proportionalibus," of which he gives three: 1, 33 (34–1?), and 39,000. The fourth, if I have interpreted his method correctly, would be equal to the square of 34, or approximately 1,100.

35. *Lect.* 1: 63ra. The estimate of less than one-seventh is based on the assumption that only six circles having a diameter equal to one quarter the circumference of a great circle can be drawn on the surface of a sphere. The surface not covered by these six circles approximates six triangles (*remanent sex trianguli*) whose combined area is more than the area of one circle. Therefore, the one circle occupied by land is offset by five other circles of water plus the triangular surface areas, giving six parts plus of water or less than one-seventh land. Via another method, Ptolemy reaches an approximation of one-eighth (*Almagestum* 2. 1 [Taliaferro, p. 34]).

36. " . . . quod navis a parte occidentali aridae incipere potest et ire super aquam sub hemisphaerio inferiori et transire sub pedibus nostris redeundo ad partem orientalem aridae." *Lect.* 1: 29va.

37. *Lect.* 1: 26va–vb, where brief mention is made of three regions; cf. Albert *Meteor.* 1. 1. 8 (B. 4: 486), who is in turn commenting on a rather

ambiguous explanation of the formation of clouds given by Aristotle *Meteor.* 1. 3. 340a24–340b31, and Pierre d'Ailly *De impressionibus aeris* (Strassburg, 1504), 2r.

38. For typical arguments supporting the existence of a sphere of fire, see Albert *Meteor.* 1. 2. 6 (B. 4: 497), and Robert the Englishman's commentary on Sacrobosco, *Sphaera* 2 (Thorndike, *The Sphaera of Sacrobosco and Its Commentators,* p. 206), where the example of laughing is mentioned.

39. *Q. s. per.* 1 (Erfurt 29ra, Valencia 42r).

40. Albert *De natura locorum* 3. 1. (B. 9: 566). Such considerations are what Albert refers to as "cosmographia."

41. *Lect.* 1: 75ra–rb; for a more detailed discussion of these zones, see Macrobius *Commentarii in Somnium Scipionis* 2. 5 (Ludovicus Ianus, *Macrobii Ambrosii Theodosii . . . Opera,* Leipzig, 1898, 1: 153–162), and Bede *De natura rerum* 9 (PL 90: 202–204).

42. *Lect.* 1: 181va–182'vb (there are two pages numbered "182"; I have added the accent to indicate the second); cf. Pliny *Naturalis historiae* 3. 1. 3 (Loeb 2: 4), and Isidore *Etym.* 14. 2. 1.

43. *Lect.* 1: 181vb, 182'rb–va; cf. the geographic descriptions set out by Pliny in *Nat. hist.* bks. 3–6, and Isidore *Etym.* bk. 14.

44. Kimble, *Geography,* pp. 208–210. Kimble's accounts of the motivation behind medieval cosmography must be read with caution. He is convinced that orthodoxy had a restraining influence on cosmographical speculations, a point that may be true, but that he frequently carries to extremes.

45. The impact of Europe's broadening horizons on its populace is discussed briefly by Edward Cheyney, *The Dawn of a New Era, 1250–1453* (New York, 1962), pp. 276–297.

46. A brief description of classifications of the sciences put forth during the Renaissance can be found in Robert Flint, *Philosophy as Scientia Scientiarum, and A History of Classifications of the Sciences* (London, 1904), especially, pp. 97–131.

47. See, for example, *Sphaera Ioannis de Sacrobosco, emendata Eliae Vineti Santonis* (Venice, 1536); or work that is partially related to *Sphaera* literature, Peter Apianus, *Cosmographiae introductio* (Paris, 1550).

48. Albert *Meteor.* 1. 1. 1 (B. 4: 478).

49. Thomas Aquinas *In libros Aristotelis Meteorologicorum [Expositio]* 4. 1 (ed. Raymundi Spiazzi [Rome, 1952], p. 645, par. 311).

50. *Lect.* 1: 63vb.

51. *Lect.* 1: 15ra; cf. Aristotle *Meteor.* 2. 3. 358a3–27, Albert *Meteor.* 2. 3. 15 (B. 4: 578), and Pierre d'Ailly *De imp. aeris* 10r.

52. *Lect.* 1: 35ra, 63vb, 84vb; cf. Aristotle *Meteor.* 1. 9. 346b16–347a13, and Albert's extended discussion of the water cycle in *Meteor.* 2. 1–3 (B. 4: 519ff.).

53. These phenomena are discussed by Aristotle in *Meteor.* 1. 10–12. 347a13ff., and Albert *Meteor.* 2. 1 (B. 4: 519ff.). Henry makes occasional reference to the watery phenomena other than rain but never discusses their origin.

54. *Lect.* 1: 29rb, 63va, 181ra; cf. Aristotle *Meteor.* 1. 13. 349a11ff., and Albert *Meteor.* 2. 2 (B. 4: 545ff.).

55. *Lect.* 1: 63vb.

56. *Lect.* 1: 61vb; cf. Albert *Meteor.* 3. 5. 1 (B. 4: 710–713).

cal circles, as taken up, for example, by Henry *Sent.* 48rff. For an interesting, although not always clear, discussion of Henry's conception of divine causality, see Alessio, "Causalita' naturale e causalita' divina," pp. 599–604.

19. *Lect.* 1: 144va. Henry suggests that "vulgata apud theologos" support God's immediate involvement in nature. "Immediate" in this sense is to be taken literally–that is, as meaning without any mediating or intervening element.

20. *Lect.* 1: 144vb.

21. It is possible to object at this point that there is an intermediary link, the world soul (*anima mundi*), that is moved by God and that in turn moves the orbs. Henry takes up this issue (*Lect.* 3: X3ra[241ra]), without fully resolving it. His argument closely follows Augustine *Civ. Dei* 7. 6 (PL 41: 199–200).

22. Henry's most extended discussion of celestial movers comes at the very end of the *Lecturae* 3: X1rb[239rb]–X9va[247va]. He had raised the problem of celestial movers earlier (*Lect.* 1: 75va–vb), but at this point left the issue unresolved, as does his source for some of his ideas on celestial movers, Augustine *De Gen. ad litt.* 2. 18 (PL 34: 279–280).

23. *Lect.* 3: X1va[239va]. This position in general follows Averroes's teachings in *De sub. orbium;* see chap. IV, nn. 11–12.

24. *Lect.* 3: X1va–vb[239va–vb].

25. *Lect.* 3: X9rb[247rb]. This seems to be the more important reason for rejecting this position.

26. John Buridan *Quaestiones super octo phisicorum libros Aristotelis* 1. 12 (trans. Clagett, *Science of Mechanics,* p. 536). See Henry's summary of this position, *Lect.* 3: X1va[239va].

27. Aristotle *Physica* 7. 5. 249b27ff. I am here employing Clagett's summary of the cumbersome phraseology used by Aristotle to express the laws of motion (Clagett, *Science of Mechanics,* pp. 425–432).

28. *Lect.* 3: X1vb[239vb]. The objection and reply raised in this instance stem from a debate between Avempace, whom Henry mentions, and Averroes over the role of resistance versus space in accounting for the temporal nature of motion. Averroes held that resistance is most important, thereby maintaining that zero resistance should give infinite velocity; whereas Avempace held that distance is the crucial factor, thus giving a body a finite velocity even in a void. See Averroes *Commentarium in physicam* 4. 71 (Venice, 1573, 4: 158r–162r); E.A. Moody's summary of this dispute in "Galileo and Avempace: The Dynamics of the Leaning Tower Experiment," *Journal of the History of Ideas* 12 (1951): 184–193; and the translations given in Grant, *Source Book,* pp. 253–263.

29. *Lect.* 3: X2vb–5vb[240vb–243vb].

30. " . . . quod intelligentiae illae sint quaedam creaturae rationales alterius generis ab angelis bonis et malis et animalibus (*sic,* animis) humanis" *Lect.* 3: X9rb[247rb].

31. *Lect.* 3: X9va[247va].

32. Such would be true of the other great opponent of astrology at Paris in Henry's day, Nicole Oresme. For a discussion of Oresme's astrology, see Lynn Thorndike, *History of Magic,* 3: 398–423, especially pp. 414–415, where he discusses the limitations Oresme puts on celestial movers, and G.W. Coopland, *Nicole Oresme and the Astrologers* (Liverpool, 1952), pp. 22–25.

33. The contemporary orientation of many of Henry's remarks on astrol-

57. The distinction of two vapors is clearly set out by Pierre d'Ailly *De imp. aeris* 2v–3r.

58. *Lect.* 1: 38rb; cf. Aristotle *Meteor.* 1. 7. 344a5–344b9, and Pierre d'Ailly, *De imp. aeris* 4v–5v. Henry has much more to say about comets in his *Quaestio de cometa,* especially 1–3 (Pruckner, pp. 89–99). In this work, however, he assumes that his readers are familiar with the standard Aristotelian account of comets and does not give any background. For the purposes of an introduction to meteorology, then, the brief discussion of comets in the *Lecturae* is actually more instructive.

59. *Lect.* 1: 38rb. For a detailed discussion of the various theories commonly used to account for comets, see Albert *Meteor.* 1. 3 (B. 4: 499ff.).

60. Aristotle *Meteor.* 1. 4. 341b36–342a4. The example given by Aristotle to clarify his explanation is that of two lamps, one below the other. If the vapors of the lower lamp are ignited by the flame of the upper lamp, the flame can be seen to pass from one to the next, much like a shooting star. This can easily be duplicated using candles and the vapor that arises from one directly after it is extinguished.

61. Aristotle *Meteor.* 1. 8. 346a16–346b9.

62. *Lect.* 1: 36vb–37ra; cf. Albert *Meteor.* 1. 2. 5 (B. 4: 495–496).

63. See, for example, Pierre d'Ailly *De imp. aeris* 5r–5v, whose theory has both celestial and sublunar aspects.

64. Aristotle *Meteor.* 2. 4. 359b27–360a34; cf. Albert *Meteor.* 3. 1. 3 (B. 4: 589).

65. *Lect.* 1: 44va. Henry has very little to say about winds. What he does say seems to concur with Albert's and Aristotle's explanations; cf. also Pierre d'Ailly *De imp. aeris* 11v–12r.

66. See, for example, Albert's discussion of the twelve winds mentioned by Seneca, *Meteor.* 3. 1. 22 (B. 4: 608–611).

67. Actually, the world does not move until all is completed on the seventh day; see chap. VI, n. 17.

68. *Lect.* 1: 62va.

69. *Lect.* 1: 29ra.

70. " . . . quod terra a principio fuit uniformis per totum in suis partibus et quod nec lapides nec minerae [mineralia?] fuerunt in ea. . . . Deus tamen non fecit montes ab initio, sed secundario post lapsum hominis, sicut et multa alia. . . . " *Lect.* 1: 29ra–rb.

71. *Lect.* 1: 29rb.

72. Ibid.

Chapter VI. Day Four: The Stars, Physics, and Astrology

1. Changes in the heavens are discussed above, chap. IV, sec. 2.

2. The relevant sections from the *Lecturae* are discussed below. The sections from the *Sentences* that have the most direct bearing on cause and effect are found in the questions on book two, particularly question one: "Utrum Peripatetici senserunt omnia alia entia a primo esse facta vel potius plurima entia non habere principium effectivum pro prima parte?" *Sent.* 89va. Similar discussions are common in *Sentence* commentary literature, as for example in Facinus de Ast's *Sententiae* 2. 2–4, 8–10 (ms. Erfurt, Wissen-

schaftliche Bibliothek der Stadt, Amplonianus F. 115, 83ra–93vb, 100ra–105vb). Facinus de Ast is of particular interest in this regard, as Henry refers to him with some frequency in his commentary on book two of the *Sentences* (see chap. I, n. 79).

3. For a discussion of the importance of astrology in Henry's scientific and science-related writings, see chap. I, sec. 2.

4. For a brief discussion of the relationship between Henry's science and his political thought, see my article, "A late Medieval *Arbor scientiarum*," pp. 268–269.

5. "Propter admirari inceperunt antiquitus homines philosophari, admiratio siquidem sensibilium effectuum et transmutationum rerum corporalium superiorum et inferiorum ad inquisitionem causarum mentem incitabat humanam scientiae et cognitionis naturaliter cupidam." *De red. eff.* 116vb.

6. Ibid.

7. "Quia libenter scire vellem modum naturalis administrationis et regiminis naturaliter agentium mundi inferioris a superiori caelestium supercaelestium[que] causarum ac utrarumque inter se concatenationem et dependentiam, quam catenam auream vocant. . . . " *De hab. caus.* 193vb. Henry's source for the term "catena aurea," as he himself notes (*Lect.* 1: 141va), is Macrobius *Commentarii in Somnium Scipionis* 1. 14. 16 (Ianus, 1: 83). For a discussion of further sources, see Arthur Lovejoy, *The Great Chain of Being* (New York, 1960), p. 63 and n. 53.

8. See *De red. eff.* 116vb, and *Prol.* 45vb–46ra, where "physica" is defined as "scientia quae causas rerum in effectibus suis et effectus a causis suis investigando considerat." Henry derives this definition from Hugh of St. Victor *Didascalicon* 1. 2 (trans. Jerome Taylor, *Didascalicon; a Medieval Guide to the Arts* [New York, 1961], p. 48).

9. Aristotle's definition of physics and its subsequent impact on medieval thinkers is discussed by Marshall Clagett, "Some General Aspects of Medieval Physics," *Isis* 39 (1948): 29–36. For a typical discussion of how various sciences relate to the study of moving bodies, see Albert the Great *Meteor.* 1. 1. 1 (B. 4: 477–478).

10. Aristotle *Physica* 3. 1. 200b10–29.

11. Aristotle *Physica* 2. 7. 198a14–22.

12. The importance of casuality in Western science is convincingly demonstrated in William Wallace's recent study, *Causality and Scientific Explanation;* see especially, 1: 1–24.

13. See the discussion of plant growth, chap. VII, sec. 1. It should be noted, as will be discussed later, that the sun is not the efficient cause of germination.

14. For Aristotle's discussion of the four causes, see *Physica* 2. 7. 198a14–198b9.

15. "Intelligentiae nullam actionem habent in istis inferioribus nisi mediantibus qualitatibus influentialibus orbium caelestium et stellarum." *De hab. caus.* 194ra.

16. *Lect.* 1: 144va, 3: X3ra[241ra].

17. *Lect.* 1: 144rb. The consequences of this analogy, which Henry fully realizes and accepts, is that technically the universe is not set into motion until the seventh day when all is completed.

18. *Lect.* 1: 144va. It should be noted that these and the remaining generalizations about God and creation were subject to question in theologi-

ogy is apparent in his references to "moderni astrologi," "doctores nostri," and so forth. For background on astrology at Paris, see Pruckner, *Studien zu den astrologischen Schriften*, pp. 73–85.

34. *De hab. caus.* 194ra, *Contra ast.* 1. 14 (Pruckner, p. 160).

35. *De hab. caus.* 194rb, *Lect.* 1: 45vb, *Contra ast.* 1. 14 (Pruckner, p. 160).

36. *De hab. caus.* 198ra, 199rb, *Contra ast.* 1. 9, 14 (Pruckner, pp. 153, 160–161).

37. " . . . ad salvandum cursum effectuum naturalium in mundo inferiori." *De red. eff.* 122va; cf. *Contra ast.* 3. 3 (Pruckner, pp. 197–198), *Lect.* 1: 60vb.

38. *De red. eff.* 121ra.

39. *De hab. caus.* 195ra–196va, *De red. eff.* 122vaff. Most of the discussion in *De reductione* is aimed at accounting for actions that others ascribe to occult agents. Henry does, of course, utilize occult forces to explain unusual phenomena.

40. " . . . in generatione plantae vel animalis, determinatio vel propria dispositio ad formam vel ad speciem non est ex aliqua imagine vel influentia caeli sed est ex intrinseca virtute seminum. . . . Caelum autem concurrit ad generationem animalium et vegetabilium tamquam agens universale excitans vel vigorans virtutes generativas seminum vel animalium in regione elementari." *Lect.* 1: 43ra. The example of plant growth is given a number of times to illustrate this point; cf. *Lect.* 1: 90vb, *Contra ast.* 3. 1 (Pruckner, p. 194).

41. *Contra ast.* 2. 4 (Pruckner, p. 177).

42. Augustine is not quoted by name in any of Henry's scientific works.

43. See Aristotle *De generatione* 1. 5. 319b7–12, for a definition of alteration.

44. *De red. eff.* 119va, 125ra–rb.

45. *De hab. caus.* 196ra.

46. *De red. eff.* 123ra; cf. 3 Kings 1: 4.

47. *Lect.* 1: 34vb.

48. *De red. eff.* 122va. A physician might, for example, prescribe a cure for a headcold (*rheuma*), that might more simply be cured by closing the windows and blocking the lunar influences that cause it (*Contra ast.* 1. 9 [Pruckner, p. 154]). For more on disease, see chap. VIII, sec. 1.

49. For a summary of other examples given in *De habitudine* and *De reductione*, see Thorndike, *History of Magic*, 3: 476–492.

50. Aristotle *De generatione* 1. 4. 219b15–22.

51. *Lect.* 1: 12rb.

52. " . . . ex combinatione et commixtione aluminis, arsenici, salisammoniaci, vitreoli, et huiusmodi rerum miscibilium et mineralium . . . secundum artificiales decoctiones." *De red. eff.* 128ra.

53. For a discussion of the generation of minerals, see Aristotle *Meteor.* 3. 6. 378a17–378b7, Albert *Meteor.* 3. 5. 1 (B. 4: 701–703), and the discussion of water above, chap. V, n. 56.

54. *De red. eff.* 123vb.

55. *Contra ast.* 3. 3 (Pruckner, pp. 197–198).

56. *De red. eff.* 123va–124rb, 127rb–vb. Henry's psychology is discussed below, chap. VIII, sec. 2.

57. *De red. eff.* 124va. Henry does not say what makes these qualities sensible. One might properly object that he is altering terms, not concepts.

58. *De red. eff.* 127vb.

59. *De red. eff.* 119va and 129ra, where Nicole Oresme's *De configurationibus qualitatum* is cited, Henry's obvious source for many of his ideas on intension and remission; see Clagett, *Nicole Oresme,* pp. 114–121.

60. *De red. eff.* 115vb.

61. See n. 59, and chap. III, sec. 1.

62. The complete title of *De habitudine causarum* continues *et influxu naturae communis respectu inferiorum. De hab. caus.* 193vb.

63. *De hab. caus.* 202vb–203ra. For a discussion of subsequent speculations about this phenomenon, see Thorndike, *History of Magic,* 3: 476–477.

64. *De hab. caus.* 200rb, 203rb–204ra.

65. The only other source of supernatural causes would be angels, who, Henry does admit, can produce local motion in the inferior region (*Lect.* 1: 87ra).

66. See *Q. s. per.* 1 (Erfurt 29rb, Valencia 47v), where natural causes are said to proceed through transmutations whereas supernatural causes proceed immediately and without transmutation; *Tractatus de discretione spirituum,* Vienna, Österreichische Nationalbibliothek, CVP 5086, 99r, which delves into the spirits that "immediately" move men as compared to the movements caused by physical forms; and the mentions of God's direct actions in nature in *Lect.* 1: 14va, 74va, 85vb, 157vb. As noted above (chap. II, sec. 3), a discussion of the actions of spirits and demons as discussed at great length in Henry's later writings is a topic that is certainly worth an extensive study.

67. Thorndike, *History of Magic,* 3: 476.

68. *Lect.* 1: 42va.

69. "Ideo mirandum, quomodo quidam beani adhuc artis astrologicae audeant se de iudiciis secundum eam sic temerarie intromittere suis mendaciis popularium mentes ad frivolas suspendo vanitates." *Contra ast.* 2. 4 (Pruckner, p. 176). For a similar criticism, see *Lect.* 1: 42rb.

70. "Odit observantes vanitates supervacuae Universitas Parisiensis." *Contra ast.* 1. 1 (Pruckner, p. 139).

71. " . . . auctoritas antiquorum, qui varios libros scripserunt de iudiciis astrorum, quorum auctoritati etiam aliqui Christianorum timeo credunt sicut equi et muli, quibus non est intellectus." *Lect.* 1: 41va.

72. "Et quamvis dicta illius vani loqui non sint digna seriosa responsione, tamen propter simpliciores quosdam qui talium superstitionum auctoribus, nesio qua levitatis aut vanitatis prurigine titillante, nimis cito adhaerere consueverunt, volo aliquid dicere contra illum rudem et indiscretum astrologicae superstitionis zelatorem." *Lect.* 1: 77va.

73. *Lect.* 1: 77vb, 84ra, *Prol.*50va.

74. See chap. IV, sec. 2.

75. *Lect.* 1: 19ra–rb, *Contra ast.* 1. 2 (Pruckner, pp. 140–141).

76. See *Lect.* 1: 42rb, for mention of the first two objections; *Lect.* 1: 36rb–va, *Contra ast.* 1. 4ff. (Pruckner, p. 144ff.), for the third; and *Lect.* 1: 42ra, for the last.

77. *Lect.* 1: 33vb.

78. *Lect.* 1: 39ra–rb; the constellations are ennumerated by Ptolemy *Almagestum* 7. 5. and 8. 1 (Taliaferro, pp. 234–258). For a rejection of such superstitions, see Isidore *Etym.* 3. 27. 2.

79. *Lect.* 1: 70ra.

80. *Lect.* 1: 70va–vb.
81. *Lect.* 1: 71ra.
82. *Lect.* 1: 71va–vb.
83. *Lect.* 1: 74rb–va; cf. Avicenna *Liber de anima* 4. 2 (ed. S. Van Riet and G. Verbeke [Louvain, 1968], pp. 16–22).
84. *Lect.* 1: 34ra–rb.
85. *Prol.* 61va–vb.
86. Thorndike, *History of Magic,* 3: 502.

Chapter VII. Day Five: Plants, Animals, and the Biological Sciences

1. " . . . anima est actus corporis physici organici in potentia vitam habentis." *Nota.* 77ra; cf. Aristotle *De anima* 2. 1. 412a28–30.
2. The lack of interest among scholastic commentators in Aristotle's treatises on animals is evident in such surveys as Charles Lohr, "Medieval Latin Aristotle Commentaries," *Traditio* 23 (1967): 313–413, and continuing.
3. This classification is given in *Nota.* 77ra; cf. Aristotle *De anima* 2. 3. 414a28–31. Henry makes clear that such powers are distinguished "by reason"; one living creature does not have as many souls as powers of the soul.
4. *Lect.* 1: 95va.
5. *Lect.* 1: 88va. If reason were involved, spider webs and beehives would take on individual characteristics, which they do not. All spider webs are the same, as are all beehives. For a discussion of this distinction, see Albert the Great *De anima* 3. 1. 3 (Stroick, 168–169), and the discussion of the internal senses, chap. VIII, sec. 2.
6. *Nota.* 77ra; cf. Aristotle *De anima* 2. 3. 414a31–414b20.
7. *Lect.* 1: 67ra.
8. *Lect.* 1: 66va.
9. *Lect.* 1: 15vb; cf. Pliny *Nat. hist.* 18. 20. 85 (Loeb 5: 245).
10. *Lect.* 1: 85vb, also 83va. Henry's discussion of putrefaction in the *Lecturae* is very similar to an earlier discussion of the same subject in *De hab. caus.* 203rb–vb. In both he stresses the fact that it is the mutual interaction of the elements with celestial influences that gives rise to worms and other creatures. It is interesting to note that in *De habitudine* no mention is made of how closely this theory follows Augustine's doctrine of *seminales;* whereas in the *Lecturae,* it is Augustine's doctrine of *seminales* that at one point leads Henry to consider putrefaction.
11. Seminal reasons are discussed at length in *Lect.* 1: 85rb–86vb. Henry's primary source for this discussion is Augustine *De trinitate* 3. 4, 8 (PL 42: 873, 875–877). Several more specific points are drawn from *De gen. ad litt.,* especially 3. 12 (PL 34: 287), 6. 11 (PL 34: 347).
12. *Lect.* 1: 85va–vb. See McKeough's discussion of these distinctions, *The Meaning of the rationes seminales,* Catholic Univ. of America, Philosophical Studies 15 (Washington D.C., 1926), pp. 35–46.
13. "Sicut de vitulorum carnibus putridis apes fiunt, et de carnibus equorum scarabei, et ita de mulis locustae, et de cancris scorpiones." *Lect.* 1: 86ra; cf. Isidore *Etym.* 11. 4. 3.
14. *Lect.* 1: 81ra–rb.

15. " . . . ipse creator, Deus, volens tot et tam varias rerum species esse ad decorem universi et gloriam suam . . . ,disposuit in eis miram partium et membrorum, interiorum et exteriorum, diversitatem et heterogenitatem atque mirandam organorum dispositum. . . . " *Lect.* 1: 99rb.

16. *Lect.* 1: 99rb–va; cf. Aristotle *De partibus animalium* 3. 3. 665a11–15, 3. 4. 665a26ff., where discussions of the importance of the heart in living creatures are contained, and my article, "A Late Medieval Debate concerning the Primary Organ of Perception," *Proceedings of the XIIIth International Congress of the History of Science* 3, 4 (Moscow, 1974): 198–204, where the rationale behind the central importance of the heart is discussed.

17. ". . . . pelles, et pilos, cornua, dentes, et ungulas, squamas, et caeteras armaturas. . . . " *Lect.* 1: 100vb.

18. " . . . adhuc magis in particulari ad ostensionem divinae potentiae et sapientiae mirabilium dispositionum in animalibus et in eorum partibus, legantur 26 libri quos de animalibus scripsit Albertus Magnus." *Lect.* 1: 100rb.

19. Augustine *De gen. ad litt.* 3. 9 (PL 34: 284) suggests that this association is made by some philosophers. For a clear elaboration of this theory, see Isidore *Etym.* 12. 3. 1–3.

20. *Lect.* 1: 97vb; cf. Augustine *Civ. Dei* 21. 2 (PL 41: 709–710).

21. *Lect.* 1: 98ra; cf. Pliny *Nat. hist.* 31. 44. 95 (Loeb 8: 436), for a discussion of the allec.

22. *Lect.* 1: 98ra.

23. *Lect.* 1: 98rb; cf. pseudo Aristotle *Problemmata* 18. 916b1–19. Henry suggests that their melancholic temperaments might have come from too much late-night study.

24. *Lect.* 1: 84rb; cf. Aristotle *Hist. animal.* 1. 1. 487b5ff. The only examples given by Henry are bats and flies. The remainder are obvious, except for the sponge, which is given by Aristotle and Albert to illustrate a similar division; see Albert, *De animalibus libri XXVI* 1. 1. 3 (ed. Hermann Stadler, 2 vols., Beiträge zur Geschichte der Philosophie des Mittelalters, vols. 15, 16 [Munich, 1916, 1921], 1: 13, 32).

25. Aristotle *Hist. animal.* 1. 5. 489a34ff; cf. Albert *De animal.* 1. 1. 6 (Stadler 1: 29, 77–79).

26. *Lect.* 1: 92vb. Henry completely glosses over the fact, mentioned by Aristotle *Hist. animal.* 1. 5. 489b10 12, that sharks are internally oviparous.

27. *Lect.* 1: 83va, 93ra; cf. Albert *De animal.* 1. 1. 6 (Stadler 1: 30–31, 81).

28. *Lect.* 1: 88rb–va, 95va, and n. 5 of this chapter.

29. *Lect.* 1: 88vb. Henry follows Albert in this regard in making man the measure of all lower animals; cf. Albert *De animal.* 2. 1. 1 (Stadler 1: 224, 1–2).

30. *Lect.* 1: 88vb. The description of the crab's internal organs may be Henry's. Aristotle *Hist. animal.* 4. 2. 525b3ff., and Albert *De animal.* 24. 1 (Stadler 2: 1527, 22), have detailed descriptions of the crab's internal organs, while Pliny *Nat. hist.* 9. 51. 97 (Loeb 3: 228), and Isidore *Etym.* 12. 6. 51, have nothing to counter their descriptions.

31. *Lect.* 1: 100va.

32. *Lect.* 1: 100va–vb.

33. *Lect.* 1: 53va.

34. Reference in this instance is to Arthur Lovejoy's *The Great Chain of*

Being. It is very difficult to place Henry within the context of the development of this idea. He simply does not bother himself with most of the metaphysical and theological problems that are, in Lovejoy's opinion, so central to this idea. When he does discuss teleology and ontology, it is not necessarily in conjunction with the great chain of being. It is interesting to note that Lovejoy includes the golden chain (p. 62) as an aspect of the great chain of being, whereas for Henry the two chains are quite distinct. The hierarchy of creatures has nothing to do with the hierarchy of causes. Just how widespread acceptance of this dichotomy was warrants further investigation.

35. For a discussion of the distinction between the golden chain and the great chain of being, see n. 34.

36. See n. 18.

37. *Contra ast.* 2. 4 (Pruckner, p. 178), *De hab. caus.* 194vb.

38. For instances in which the magnet is used as an illustrative example, see *Lect.* 1: 34vb, 43va, 70rb, 72rb, 73vb, *Contra ast.* 1. 10, 2. 4, 3. 4 (Pruckner, pp. 154, 177–178, 203), *De red. eff.* 124va. Despite his claim that magnets do not operate through occult forces, Henry constantly compares their operations to the operation of occult forces.

39. *Contra ast.* 2. 4 (Pruckner, p. 178).

40. *Lect.* 1: 31rb; cf. Pliny *Nat. hist.* 19. 44. 155 (Loeb 5: 520, and n. b).

41. *Lect.* 1: 31va; cf. Ambrose *Hexaem.* 3. 8. 36 (CSEL 32: 83).

42. *Contra ast.* 3. 1 (Pruckner, p. 193); cf. Pliny *Nat. hist.* 20. 93. 252 (Loeb 6: 146), who assigns the same property to blite (*blitum*).

43. *De red. eff.* 124vb.

44. *Lect.* 1: 65va.

45. *Lect.* 1: 31rb–va; cf. Ambrose *Hexaem.* 3. 15. 64 (CSEL 32: 104–105).

46. *Lect.* 1: 81ra, 82ra.

47. *Lect.* 1: 80ra, 82ra.

48. *Lect.* 1: 79vb. Pliny notes, *Nat. hist.* 23. 4. 10 (Loeb 8: 470), that one large sea monster, washed up in the mouth of a river in Arabia, measured 600 feet long and 360 feet wide. Using a figure of about 40,000 square feet per acre, this would come out to about four acres.

49. *Lect.* 1: 79vb; cf. Isidore *Etym.* 8. 11. 27–28; 12. 6. 6–8.

50. *Lect.* 1: 80rb.

51. *Lect.* 1: 79ra; cf. Isidore *Etym.* 12. 4. 3; 12. 6. 2. Aristotle *Hist. animal.* 6. 2. 690b11ff. and Albert *De animal.* 14. 2. 5 (Stadler 2: 975, 54–55) give more exacting descriptions of reptiles.

52. *Lect.* 1: 96va.

53. *Lect.* 2: 1ra. The following discussion of snakes is set out in conjunction with the temptation of Eve.

54. *Lect.* 2: 1rb; cf. Isidore *Etym.* 12. 4. 12, Albert *De animal.* 25. 2 (Stadler 2: 1557–1558, 13). The asp is an Egyptian snake that looks like a cobra. It is extremely poisonous.

55. *Lect.* 2: 1va; cf. Isidore *Etym.* 12. 4. 18, Albert *De animal.* 25. 2 (Stadler 2: 1563, 17). The cerastes is an African, horned viper that has two horns, not eight, and is extremely poisonous.

56. Lect. 2: 1va–2ra; cf. Albert *De animal.* 25. 2 (Stadler 2: 1563–1570, 22, 24, 25, 30, 33). These are a mixture of water snakes and snakes that inhabit fruit trees.

57. *Lect.* 1: 83rb; cf. Isidore *Etym.* 12. 6. 19–21, Pliny *Nat. hist.* 8. 37. 89; 8. 39. 95 (Loeb 3: 64, 68). Assuming a cubit is about eighteen inches, the estimate for the size of the crocodile is reasonable; some species grow to over thirty feet in length.

58. *Lect.* 1: 87va. This classification follows Albert's discussion of birds, *De animal.* 23. 1 (Stadler 2: 1430, 1ff.). Isidore *Etym.* 12. 8. 1. classifies insects as "minuta volatilia," that is to say, small flying creatures.

59. *Lect.* 1: 93rb; cf. Albert *De animal.* 23. 1 (Stadler 2: 1431, 3).

60. *Lect.* 1: 93ra.

61. *Lect.* 1: 87vb.

62. *Lect.* 1: 103rb, where Henry refers to this example as set out in more detail by Ambrose *Hexaem.* 5. 15. 50 (CSEL 32: 178).

63. *Lect.* 1: 104rb.

64. *Lect.* 1: 95vb, 97rb–va; cf. Pliny *Nat. hist.* 8. 1. 1ff. (Loeb 3: 2ff.), Albert *De animal.* 22. 2. 1 (Stadler 2: 1376, 50). Henry does not mention, as does Albert, that elephants are afraid of mice and the grunts of pigs.

65. *Lect.* 1: 108ra–vb.

66. *Lect.* 1: 103vb–104ra; cf. Albert *De animal.* 8. 1. 3 (Stadler 1: 581, 27).

67. *Lect.* 1: 87vb.

68. " . . . cattos contra mures, araneas contra muscas, canes contra lupos." *Lect.* 1: 82ra. Henry calls such animals "correctores."

69. *Lect.* 1: 90ra.

70. *Lect.* 1: 94va; cf. Augustine *Civ. Dei* 2. 23 (PL 41: 71).

71. *Lect.* 2: 17ra.

72. *Lect.* 1: 95rb.

73. *Lect.* 1: 102ra.

74. Thorndike, *History of Magic,* 3: 502.

75. William Gilbert, *De magnete magneticisque corporibus et de magno magnete tellure physiologia nova* 1. 1, 15 (trans. P. Mottelay, *Great Books of the Western World* [Chicago, 1952], 28: 3, 20).

Chapter VIII. Day Six: Humanity and the Human Sciences

1. "Cum homo, ut dictum est, sit finis et causa creaturarum inferioris mundi." *Lect.* 1: 110ra.

2. *Lect.* 1: 110rb.

3. For a detailed discussion of the importance of this passage, see Charles Trinkaus, *In Our Image and Likeness* (Chicago, 1970), especially, 1: 18–28.

4. Compare, for example, the texts required for the arts degree (*Chart.* 2: 678) and those required for medicine (*Chart.* 1: 516–518).

5. *Prol.* 54va; see Steneck, "*Arbor scientiarum,*" p. 256.

6. " . . . quomodo non mediocriter laudabilis est et digna illa naturalis philosophia, atque illa particularis naturalium scientia, quae medicina appellatur, ex eo quod nedum vegetabilium rerum et caeterarum vires investiget, sed inventis viribus multis ingeniis morbos novit curare et sanitatem conservare." *Lect.* 1: 30vb. See also *Lect.* 1: 71ra, where Henry rates medicine above astrology as a prognoscative science, and 42va.

7. *Lect.* 1: 30vb–31ra; see chap. VI, n. 85, for Henry's comments on astrology's place in society.

8. *Lect.* 1: 88vb, *De red. eff.* 128rb–va. How these organs function is discussed below.

9. *Lect.* 1: 125rb. The theory of the four humors is at least as old as the collection of ancient Greek medical texts known as the Hippocratic corpus (in the tradition of Hippocrates of Cos, fifth century b.c.) and was commonly taken up by most medical writers thereafter. For the Hippocratic theory, see *On the Nature of Man* 4–5 (trans. W.H.S. Jones in *A Source Book in Greek Science,* ed. Morris Cohen and I.E. Drabkin [Cambridge, Mass., 1958], pp. 488–489). Henry could have derived this theory from any one of a number of sources, such as Albert *De animal.* 20. 1. 11 (Stadler 2: 1304, 59), which he uses in other contexts, or Isidore *Etym.* 4. 5. 1–10.

10. *Lect.* 2: 164va. Again, Henry could have come upon this very common division of spirits in any one of a number of sources; cf. Albert *De animal.* 1. 2. 20 and 20. 1. 7 (Stadler 1: 135, 381 and 2: 1292, 37).

11. For brief descriptions of Galenic physiology, see Charles Singer, *Greek Biology and Greek Medicine* (Oxford, 1922), pp. 66–70; Alastair Crombie, *Medieval and Early Modern Science* (Garden City, N. Y., 1959), 1: 162–168; and the translations of primary sources in Cohen and Drabkin, *Source Book,* pp. 467–486, and Grant, *Source Book,* pp. 705–715.

12. Ibid.; cf. Isidore *Etym.* 4. 5. 4–10. The association of the four humors with the four elements was frequently represented schematically as follows:

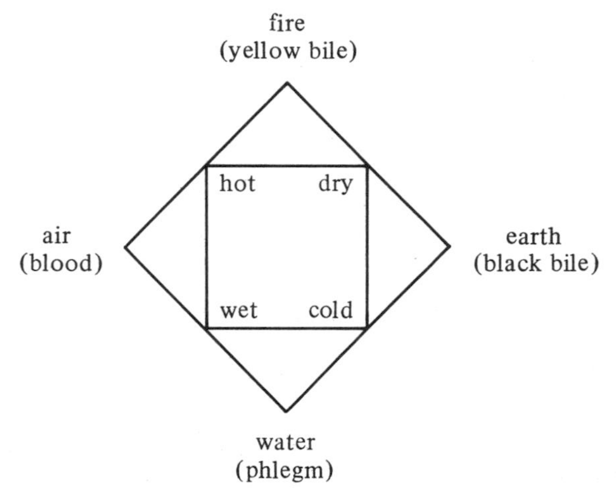

13. *Lect.* 2: 164va. The description of the animal spirit is put forth "ad considerationem medici." How animal spirit differs from the spirit of the two souls is not explained.

14. *Lect.* 1: 182vb. This suggestion is also put forth "secundum medicos" (see n. 13), in this case as an example of many rivers (the rivers of the world) having their origin in one font (the river of Paradise).

15. "Spiritus enim naturales, qui vehicula sunt virtutum seu virium animae, ex saguinis generantur humore, nec potest esse quod vectum non sequatur in multis motum sui vectoris. . . ." *Lect.* 1: 168va.

16. *Lect.* 1: 168va–vb.
17. *Lect.* 1: 168vb.
18. Further details regarding the psychological aspect of this system are discussed below, sec. 2.
19. *Lect.* 1: 170ra; cf. Aristotle *Metaphysica* 12. 6. 1072a9–18, Albert *De animal.* 1. 1. 5. and 12. 1. 1. (Stadler 1: 27, 73 and 803, 15).
20. *Lect.* 1: 170rb; cf. Albert *De animal.* 1. 1. 5 (Stadler 1: 27, 73).
21. *Lect.* 1: 170va.
22. *Lect.* 1: 166ra. Henry's reluctance to pursue a detailed description of animals is discussed in chap. VII, sec. 1.
23. *Lect.* 1: 167vb.
24. *Lect.* 1: 169va.
25. *Lect.* 1: 169va–vb; cf. Vitruvius *De architectura libri decem* 3. 1., as described by Leonardo da Vinci in Erwin Panofsky, *The Codex Huygens and Leonardo da Vinci's Art Theory* (Westport, Conn., 1971), pp. 19–23.
26. *Lect.* 1: 169vb, cf. Augustine *Civ. Dei* 15. 26 (PL 41: 472).
27. *De red. eff.* 119va–vb. Henry's discussion of disease in *De reductione* is transcribed and translated by Clagett, *Nicole Oresme,* p. 116.
28. *Contra ast.* 3. 1 (Pruckner, p. 193).
29. *Lect.* 1: 45ra, *Contra ast.* 3. 1 (Pruckner, p. 193); see also the discussion of snakes above, chap. VII, sec. 2.
30. *Contra ast.* 2. 4 (Pruckner, p. 179).
31. *Lect.* 1: 45ra, *Contra ast.* 1. 16 (Pruckner, pp. 165–167).
32. Henry begins his *Epistola . . . ad episcopum Wormaciensem* (ca. 1384), "Reverendo in Christo patri ac domino . . . Henricus . . . de Hassia, in medio regni pestilenciae. . . . " Ed. Gustav Sommerfeldt, "Die Prophetien der hl. *Hildegard von Bingen* in einem Schreiben des Magisters Heinrich v. Langenstein (1383), und Langensteins Trostbrief über den Tod eines Bruders des Wormser Bischofs Eckard von Ders (um 1384)," *Historisches Jahrbuch* 30 (1909): 298.
33. *Contra ast.* 3. 3 (Pruckner, pp. 197–199).
34. *De red. eff.* 119vb–120rb. The six proportions can be derived by adding to the four side proportions of the elemental square (see n. 12) the two diagonal proportions: hot-cold and wet-dry.
35. Oresme represented the two variable qualities with geometric figures; intensity making up a series of verticle lines that are extended (extension) over a particular subject or time, in the case of a body that is moving locally. If the intensity does not vary over the extension of the subject, then all of the verticle lines are the same and the quality is represented as a rectangular configuration. See Clagett, *Nicole Oresme,* p. 15ff. for a more detailed description of the doctrine of configurations.
36. *De red. eff.* 119vb; see Clagett, *Nicole Oresme,* p. 116.
37. Ibid., the italics are mine.
38. See note 35, for an explanation of geometric configuration.
39. *De red. eff.* 119vb; see Clagett, *Nicole Oresme,* p. 116.
40. *De red. eff.* 128ra.
41. *Lect.* 1: 34vb.
42. Thorndike, *History of Magic,* 3: 490; Clagett, *Nicole Oresme,* p. 121.
43. The roll of cause and effect in Henry's science is discussed in chap. VI, sec. 1, and considered in general terms in chap. IX, sec. 2.

44. For a general survey of the history of medieval psychology, see George Brett, *A History of Psychology* (New York, 1921), 2: 69–137. The importance of psychological writings throughout the Middle Ages is well illustrated by Jozef de Raedemaeker's survey of *De anima* commentaries in "Une ébauche de catalogue des commentaires sur le 'De anima' parus aux XIIIe, XIVe, et XVe siècles," *Bulletin de philosophie médiévale* 5 (1963): 149–183; 6 (1964): 119–134.

45. These distinctions between the approaches to psychology were not maintained by medieval psychologists. In fact, psychology, as a separate discipline, was not part of the *arbor scientiarum*, most likely since it had bearing on so many sciences. This does not mean, however, that some medieval psychologists did not lean more to one approach than another. That they did is frequently overlooked by modern scholars who tend to assume that psychology and epistemology are interchangeable and that the approach of all scholastics to the knowing process is much the same. This has led to oversights in interpreting the thought of medieval psychologists, as I show, for example, in my article, "Albert the Great on the Classification and Localization of the Internal Senses," *Isis* 65 (1974): 194–196.

46. *Lect.* 1: 70ra.

47. One aspect of the physical process involved, the multiplication of species, is traced through various fourteenth-century writers by Anneliese Maier, "Das Problem der 'species sensibiles in medio' und die neue Naturphilosophie des 14. Jahrhunderts," in *Ausgehendes Mittelalter* (Rome, 1967), 2: 419–451.

48. For a discussion of the concept of intuitive cognition, particularly as it relates to the earlier epistemological theories of Aquinas and Scotus, see Sebastian Day, *Intuitive Cognition: A Key to the Significance of the Later Scholastics,* Franciscan Institute Publications, Philosophy Series no. 4 (St. Bonaventure, N.Y., 1947). Additional references on this subject are given by Courtenay, "Nominalism and Late Medieval Thought," pp. 723–724.

49. *Lect.* 2: 3ra. Scotus's ideas on intuitive and abstractive cognition are discussed by Charles Brampton, "Scotus, Ockham and the Theory of Intuitive Cognition," *Antonianum* 40 (1965): 449–466.

50. *Lect.* 2: 114va, and *Q. s. per.* 11 (Erfurt 36vb, the printed edition of the *Quaestiones* lacks question eleven).

51. The history of the emission theory of vision is discussed by David Lindberg, "Alhazen's Theory of Vision and Its Reception in the West," *Isis* 58 (1967): 321–341.

52. Henry's clearest description of this process is given in *Nota.* 80rbff., where the sensitive power of the animal soul is discussed. However, since this explanation of sensation was commonly accepted as faithful to the text of Aristotle and hence to "common philosophy," he does not define terms or give a very detailed description. Such is also true in the *Lecturae,* where similar assumptions are made without explanation; see *Lect.* 1: 60va, 71va; cf. Aristotle *De anima* 2. 5–6. 416b32–418a6.

53. "Sensibile proprium est illud quod non contingit altero sensu sentire quam uno, sicut color sentitur visu et non alio." *Nota.* 81va; cf. Aristotle *De anima* 2. 6. 418a8–11, Albert *De anima* 2. 3. 5 (Stroick, pp. 102–104).

54. *Nota.* 83vb–84ra.

55. *De red. eff.* 123vb–124rb.

56. *De red. eff.* 123vb, see also *Nota.* 89vb; cf. Aristotle *De anima* 2. 11. 422b23–33, Albert *De anima* 2. 3. 30 (Stroick, p. 141).

57. *Nota.* 89vb; cf. Aristotle *De anima* 2. 11. 422b17–20, Albert *De anima* 2. 3. 30 (Stroick, pp. 141–142). This and the subsequent problems regarding the senses were commonly discussed in *De anima* commentaries throughout the thirteenth and fourteenth centuries.

58. *De red. eff.* 123vb.

59. *Nota.* 89vb. The question of the nobility of the senses is discussed by David Lindberg and Nicholas Steneck, "The Sense of Vision and the Origins of Modern Science," in *Science, Medicine and Society in the Renaissance*, ed. Allen G. Debus (New York, 1972), 1: 33–39.

60. *Nota.* 86ra, 87vb, 89vb; cf. Aristotle *De anima* 2. 8–10. 419b4–422b17.

61. *Nota.* 86ra; cf. Aristotle *De anima* 2. 8. 419b5–8, where wool and sponges are mentioned.

62. For surveys of the history of the internal senses, see Harry Wolfson, "The Internal Senses in Latin, Arabic, and Hebrew Philosophical Texts," *Harvard Theological Review* 28 (1935): 69–133; George Klubertanz, *The Discursive Power: Sources and Doctrine of the 'Vis Cogitativa' According to St. Thomas Aquinas* (St. Louis, 1952); and Nicholas Steneck, "The Problem of the Internal Senses in the Fourteenth Century," Ph.D. Diss., the University of Wisconsin, 1970.

63. "... continue a cerebro fluentes usque vel versus exteriora et postea quodam motu refluentes ... ad communem sensum vel ad imaginatam." *Q. s. per.* 11 (Erfurt, 36vb). Common sense and imagination are discussed below.

64. "Ex quo sequitur quod in concavitatibus nervorum deferentium spiritus sensitivos descendentium a cerebro oportet esse corpus dyaphanum ad multiplicationem specierum aptum illuminatum quod corpus terminatur ad organum exteriorum sensuum. . . ." *Q. s. per.* 11 (Erfurt, 36vb).

65. *Nota.* 94vb; cf. Avicenna *Liber de anima seu sextus de naturalibus* 1. 5 (Louvain, 1968, 87–89), Albert *De anima* 2. 4. 7 (Stroick, p. 136), Aquinas *Sum. theol.* 1a. 78. 4 (BF 11: 134–143).

66. *Nota.* 81va, 94vb; cf. Albert *De anima* 2. 4. 8–12 (Stroick, pp. 158–165), Aristotle *De anima* 3. 1. 425a14–20, 27–30.

67. *Nota.* 94vb. For an example of the detail that could be put forth in the discussion of these powers, see Albert *Summae de creaturis* 2. 35–42 (B. 35: 306–361), and my discussion of Albert's treatment of the internal senses in "Albert the Great," pp. 197–211.

68. For the issues involved in discussions regarding whether the heart or brain is the primary organ of perception, see my article, "A Late Medieval Debate Concerning the Primary Organ of Perception," pp. 198–200.

69. *Nota.* 95 ra; cf. Albert *De anima* 2. 4. 7 (Stroick, p. 158).

70. *Nota.* 95ra.

71. See Albert *Summae de creaturis* 2. 38. 3 (B. 35: 328).

72. Apart from his commentary on *De anima,* Henry discusses the operation of the external senses in *De red. eff.* 123vb–124rb.

73. Aristotle *De anima* 3. 1–3. 424b20–429a9.

74. Brief reference to the internal senses is made in *Q. s. per.* 11 (Erfurt, 36vb), and *Lect.* 1: 13vb, 54vb, 71ra, where "imaginativa, memorativa, [et] cogitativa" are listed, and 71va.

75. Aristotle *De anima* 3. 3. 427a16–429a9.
76. *Nota.* 97ra.
77. *Nota.* 97rb; cf. Aristotle *De anima* 3. 3. 428b30–429a1.

Chapter IX. Day Seven: Rest and Reflection

1. *Lect.* 1: 141ra–142rb.
2. Particularly important in this regard would be a comprehensive analysis of the scientific writings of some of Henry's colleagues while in the arts at Paris and at the Court of Charles V, especially: Dominicus of Chivasso, Evrard of Conty, Gerhard Groot, Henry of Oyta, James of Eltville, John of Chymiacho, John of Basel, John of Wasia, Marsilius of Inghen, Pierre d'Ailly, and most importantly, Nicole Oresme.
3. At times Henry directly equates the "common philosophy" with statements made by Aristotle, as for example *Lect.* 1: 23rb, 2: 4ra.
4. For a discussion of the stress placed on Augustine's writings in the fourteenth century, see Damascus Trapp, "Augustinian Theology of the 14th Century," *Augustiniana,* especially pp. 146–197.
5. *Lect.* 1: 19va–vb. Aquinas's and Augustine's views on the importance of the study of nature to theology are discussed by William Wallace, *Causality and Scientific Explanation,* pp. 71–72, and Frederick Copleston, *A History of Philosophy* (New York, 1962), 2. 1: 86. It should be noted that at one point (*Lect.* 1: 9va) Henry uses Aquinas's opinions (*Summa contra gentiles* 2. 2 [trans. English Dominican Fathers, London, 1923, 2: 2–4]) to support this position and not Augustine's. However, Henry does not go on, as does Aquinas, to distinguish between the way philosophers and theologians treat creatures (Aquinas *Sum. contra gen.* 2. 4 [DF 2: 6–8]).
6. See chap. I, sec. 3.
7. This point is discussed in chap. I, sec. 3.
8. The contributions of the English "calculatores" and Ockhamism—a term that implies many things, none of which seem to apply to Henry's thought—have been described in numerous secondary works. In addition to the material given by Duhem and Maier, see James A. Weisheipl, "Ockham and Some Mertonians," *Mediaeval Studies* 30 (1968): 163–213; William Courtenay, "Nominalism and Late Medieval Thought," especially pp. 726–730, for literature on Ockhamism; and Wallace, *Causality and Scientific Explanation,* pp. 53–64.
9. Thorndike, *History of Magic,* 3: 476–477, suggests that there are similarities between Henry's and Bacon's ideas on common natures, without speculating as to how Henry may have come across this idea. Henry upon rare occasion cites William of Auvergne's *De universo creaturarum* (*Lect.* 1: 40va, 100rb) with regard to some crucial ideas on celestial causality.
10. Among the many works that discuss the significance of the methodology employed by medieval scientists, see Alastair Crombie, *Robert Grosseteste and the Origins of Experimental Science* (Oxford, 1953); Edward Grant, "Hypotheses in Late Medieval and Early Modern Science," *Daedalus* 91 (1962): 599–616; Pierre Duhem, *To Save the Phenomena,* trans. E. Doland and C. Maschler (Chicago, 1969); and Amos Funkenstein, "The Dialectical Preparation of Scientific Revolutions," in *The Copernican Achievement,* ed. Robert S. Westman (forthcoming).

11. *Lect.* 1: 41rb; cf. Avicenna *Liber canonis* 2. 1. 1–4 (Venice, 1507, fols. 81rb–86rb).

12. The outlines of this method as well as practical examples of how it is applied are most clearly given in *De habitudine* and *De reductione*, see chap. VI, nn. 5, 7.

13. In addition to the works cited above, n. 10, see John Herman Randall, *The School of Padua and the Emergence of Modern Science* (Padua, 1961), for a discussion of the links between Renaissance and modern science.

14. (a) *Lect.* 1: 22ra, (b) *Lect.* 1: 52va, (c) *Lect.* 1: 47va, (d) *Lect.* 1: 49ra, (e) *Lect.* 1: 58va, (f) *Lect.* 1: 74va, (g) *Lect.* 1: 77va.

15. *Lect.* 1: 74va.

16. *Lect.* 1: 50va.

17. Examples of these three uses of experience can be found in *Lect.* 1: 62va, *Lect.* 1: 61va, and *Q. s. per.* 1 (Erfurt 30ra, Valencia 48v). For a discussion of the importance of arguments "secundum imaginationem," see Oresme, *De configurationibus* 1. 1 (Clagett, *Nicole Oresme*, pp. 164–167).

18. *De hab. caus.* 194ra.

19. *Lect.* 1: 23va.

20. *Lect.* 1: 88vb.

21. Henry returns to this topic, which is discussed at some length in *Prologus*, in his *Sermo*; see especially, pp. 143–146.

22. For an interesting statement on the unity of truth, see Augustine *De doctrina Christiana* 4. 29 (PL 34: 119–120).

23. See my discussion of Henry's classification of the sciences in "A Late Medieval *Arbor scientiarum*," pp. 245–269.

24. These ideas are most clearly set out in the context of Henry's discussion of how one ought to study Scripture (*Lect.* 1: 10va–11ra), and of whether science and theology teach contrary truths (*Sent.* 9ra–22ra).

25. *De hab. caus.* 195rb.

26. *Q. s. per.* 5 (Erfurt 32va, Valencia 53r).

27. *De hab. caus.* 194rb, 196ra.

28. *Contra ast.* 3. 3 (Pruckner, p. 198).

29. *Q. s. per.* 1 (Erfurt 29rb, Valencia 47v).

30. *Q. s. per.* 1 (Erfurt 29ra, Valencia 47r).

31. The case for the applicability of the methods used in the *Lecturae* to the pursuit of science in general could be further strengthened by noting that historical considerations, and the procedures stemming from these, had a well defined place in Aristotelian commentaries. Aristotle himself sorted through prior developments relating to specific sciences in order to strengthen and clarify his own arguments (Aristotle *Physica* 1. 2–4. 184b15–188a17, *Metaphysica* 1. 3–9. 983a24–993a10). Moreover, a great deal of what is properly considered by modern historians to be science was carried out within the context of theological treatises (for examples, see Grant, *Source Book*, pp. 316–324). This case must, of course, not be carried to extremes. Theological considerations are obviously more important in the *Lecturae* than in the related scientific works.

32. For an example of the application of this method to science, see my article, "A Late Medieval Debate concerning the Primary Organ of Perception," pp. 198–204. The allusion to cathedral building is obviously drawn from Erwin Panofsky, *Gothic Architecture and Scholasticism* (New York, 1957).

33. These examples of limitations that Henry puts on knowledge are given in *Lect.* 1: 18va, 22va, *De red. eff.* 128rb, *Lect.* 1: 96vb, 134ra.

34. Probabilism and hypothetical argumentation are usually thought to be important features of fourteenth-century methodology, particularly as set out by Buridan; see Wallace, *Causality and Scientific Explanation,* pp. 107–108, and Theodore Scott, "John Buridan on the Objects of Demonstrative Science," *Speculum* 40 (1965): 654–673. Since there is essentially no theoretical basis to Henry's tentative solutions, there is no reason why he could not be following the lead of Augustine, who recommended caution in reaching conclusions on difficult matters (Augustine *De gen. ad litt.* 2. 18 [PL 34: 279–280]).

35. *Lect.* 1: 38va.

36. *Lect.* 1: 35va.

37. In this regard, the reader must keep in mind that the object of the discussion that follows is science (see chap. I, n. 3) and not philosophy. I am perfectly willing to grant that there are philosophical differences that separate fourteenth-century Aristotelian naturalism from seventeenth-century mechanical philosophy. My purpose in pursuing this analysis is not to deny such philosophical differences but to point out that on the more material level of actual descriptions of nature, and considering the way the ingredients of world views operate, there is a great deal of similarity between the way a matter-form world and a world that is composed of atoms functions, even if specific details relating to the way they function are different. The implications this observation has for interpretations of the Scientific Revolution obviously falls outside the domain of this study. However, it should be noted that historians of science are coming to realize that major shifts in scientific thinking are not necessarily connected with complete changes in world views. Major "themes," to use Holton's terminology, do persist even through major scientific advances (Gerald Holton, *Thematic Origins of Scientific Thought* [Cambridge, Mass., 1973]).

38. R.G. Collingwood, *The Idea of Nature,* p. 5.

39. Henry's clearest description of God's role in nature and the origins of the laws of nature comes at the very end of the *Lecturae* 3: X3ra–vb [241ra–vb].

40. See, for example, Kenneth Dougherty, *Cosmology: An Introduction to the Thomistic Philosophy of Nature* (Peekskill, N.Y., 1953).

Bibliography

Primary Sources

1. Manuscript copies of the *Lecturae super prologum et Genesim**

Admont, Stiftsbibliothek 154, 480 fols.

Berlin, Deutsche Staatsbibliothek, Lat. 2° 709, fols. 1–104.

Dresden, Koniglich Öffentlichen Bibliothek P.28, fols. 1–251.

Erfurt, Wissenschaftliche Bibliothek der Stadt, Amplonianus F. 56, fols. 1–44; F. 69, 298 fols; F. 140, 241 fols.

Gdańsk, Biblioteka Gdańska Polskiej Akademii Nauk 2° 237, fols. 1–129; 4° 22, fols. 1–11 (*excerpta*).

Giessen, Universitäts-Bibliothek 779 (Butzbach, BG. XV. 1), fols. 1–323.

Göttweig, Stiftsbibliothek XV. 204, 384 fols.

Hamburg, Staats- and Universitäts-Bibliothek, Theol. 1009.

Heiligenkreuz, Stiftsbibliothek, Cistercienser 155, fols. 118–125 (*excerpta*).

Karlsruhe, Badische Landesbibliothek, Reichenauer. 41, fols. 1–138; 42, fols. 1–138; 43, fols. 1–159; 44, fols. 1–166; 45, fols. 1–245; 46, fols. 1–112; 47, fols. 1–128; 109, fols. 102–105 (*excerpta*).

Klagenfurt, Bibliothek zu Klagenfurt, Pap. Hs. 29, fols. 74v–75v (*excerpta*).

Klosterneuburg, Stiftsbibliothek 332, fols. 1–152; 333, fols. 1–331; 334, fols. 1–168; 335, fols. 1–213; 492, fols. 1–191; 700/4, fols. 145–288.

Kopenhagen, Koniglige Bibliothek, Ny. Kgl. Saml. 4° 111, fols. 6–14 (*excerpta*).

Kraków, Biblioteka Jagiellońska 1359, fols. 21–345; 1360, fols. 1–220; 2185, fols. 1–222.

Lambach, Stiftsbibliothek, Cod. cart. 7, fols. 1–241; 15, 232 fols.; 17, 249 fols.

Linz, Studienbibliothek, Priesterseminar II. 5, fols. 16–221.

Maihingen (Schloss Harburg), Fürstlich Bibliothek, II lat. 1. 2° 26, fols. 1–133.

Melk, Stiftsbibliothek, 34.[A.40], 328 fols; 591[L.10.], 371 fols.

Munich, Bayrische Staatsbibliothek, CLM 4403, fols. 244–247; 5196, fols. 1–118; 7507, fols. 85–191; 12717, fols. 1–117; 18145, fols. 1–337;

*This list of manuscripts is derived from Stegmüller and a search of catalogues for manuscript collections not included in *Repertorium Biblicum Medii Aevi,* 3: 31–34. It is included to illustrate the popularity of the *Lecturae* and is not intended to provide a basis for a critical edition. To date, I have not examined all of the copies of the *Lecturae* in person and so cannot at this point present such a critical listing.

18146, fols. 1–497; 18147, fols. 1–433; 18464, fols. 174–187; 18521, fols. 1–119; 18647, fols. 77–81 (*excerpta*); 19610, fols. 59–82; 26608, fols. 88–302.
Munich, Universitätsbibliothek, 2° 73, fols. 209–318.
Nuremberg, Stadtbibliothek, Cent. II. 10, fols. 1–199; Cent. II. 11, fols. 1–156; Cent. IV. 5, fols. 1–155.
Oxford, Lincoln College LXII, fols. 1–122.
St. Florian, Stiftsbibliothek, XI.21, fols. 1–341.
St. Paul, Stiftsbibliothek, Cod. Hosp. chart. 140, no. 9.
Seitenstetten, Stiftsbibliothek, Cod. CXIV, entire.
Stuggart, Württembergichen Landesbibliothek, HB. VI. 130, fols. 31–166.
Vatican, Biblioteca Apostolica Vaticana, Barb. Lat. 611.
Vienna, Österreichische Nationalbibliothek, CVP 3737, fols. 212v–222v; 3900, fols. 2r–395v; 3901, fols. 1–350; 3902, fols. 1–254; 3919, fols. 1–406; 3922, fols. 1–443; 4379, fols. 1–418; 4380, fols. 1–346; 4423, fols. 1–185; 4446, fols. 1–450; 4571, fols. 7r–108r; 4638, fols. 1–198; 4660, fols. 1–200; 4646, fols. 1–432; 4652, fols. 1–217; 4825, fols. 1–130; 4816, fols. 1–99; 4861, fols. 1–228; 4651, fols. 1–127; 4678, fols. 1–182; 4677, fols. 1–260; 4679, fols. 1–220; 4821, fols. 1–124; 4830, fols. 1–118.
Vienna, Dominikanerkonvent, Cod. 155/125(b), fols. 419–427; Cod. 155/125(a), fols. 1–415.
Vienna, Schottenkloster 112–101, fols. 1–331; 330–310, fols. 1–398.
Wolfenbüttel, Cod. Guelf, 81. 20. Aug. 2°, fols. 1–333; 81. 21. Aug. 2°, fols. 1–221; 81. 22. Aug. 2°, fols. 1–304.
Wrocaw (Breslau), Biblioteka Uniwersytecka 91 (I.F.86), fols. 1–113; 92 (I.F.87), fols.

2. Other works by Henry of Langenstein

Henry of Langenstein. *De habitudine causarum et influxu naturae communis.* Mss. Brussels, Bibliothèque Royale 21856, fols. 93r–103v; Erfurt, Wissenschaftliche Bibliothek der Stadt, Amplonianus Q. 298, fols. 68r–84v; Florence, Biblioteca Medicea Laurenziana, Ashburnham 210(142), fols. 145r–158v; London, British Museum, Sloane 2156, fols. 193vb–208va; Paris, Bibliothèque Nationale, Latin Ms. 14887, fols. 42v–65v, Latin Ms. 16401, fols. 68r–91r; Paris, Bibliothèque de l'Université 582, fols. 1r–10v; Vatican, Biblioteca Apostolica Vaticana, Vat. Lat. 3088, fols. 14r–26r; Vienna, Österreichische Nationalbibliothek, CVP 4217, fols. 1r–9v; Wurzburg, Universitätsbibliothek 292, fols. 13r–24v.
———. *De medicinis simplicibus.* Mss. Munich, Bayrische Staatsbibliothek, CLM 3073, fols. 247ra–282ra; Vatican, Biblioteca Apostolica Vaticana, Pal. Lat. 1279, fols. 109ra–119rb.
———. *"de motibus planetarum secundum eccentricos et epicyclos."* (lost, see chap. I, n. 31).

———. *De reductione effectuum*. Mss. Erfurt, Wissenschaftliche Bibliothek der Stadt, Amplonianus Q.150, fols. 25r–46r; Amplonianus Q. 298, fols. 85v–97v; Florence, Biblioteca Medicea Laurenziana, Ashburnham 210(142), fols. 89rb–101ra; London, British Museum, Sloane 2156, fols. 116vb–130va; Paris, Bibliothèque de l'Arsenal 522, fols. 57r–65v, 2128, fols. 77r–88r; Paris, Bibliothèque Nationale, Latin Ms. 2831, fols. 103r–115v, Latin Ms. 14580, fols. 205r–213r, Latin Ms. 14887, fols. 65v–88r, Latin Ms. 16401, fols. 92r–106r; Vienna, Österreichische Nationalbibliothek, CVP 4217, fols. 29r–38r.

———. *De reprobatione eccentricorum et epicyclorum*. Mss. Melk, Stiftsbibliothek 51, fols. 210r–218r; Oxford, Bodleian Library 300, item 5; Paris, Bibliothèque Nationale, Latin Ms. 16401, fols. 55r–67v; Prague, Knihovna Metropolitni Kapituli 1272, fols. 45r–54r; Princeton, Princeton University Library, Garret 95, fols. 146r–167v; Utrecht, Bibliotheek der Universiteit 725, fols. 218r–246r; Vatican, Biblioteca Apostolica Vaticana, Vat. Lat. 4082, fols. 87r–97r; Vienna, Österreichische Nationalbibliothek, CVP 5203, fols. 100r–117v.

———. *Dici de omni*. Mss. Erfurt, Wissenschaftliche Bibliothek der Stadt, Amplonianus Q. 150, fols. 93r–104v, Klosterneuburg, Stiftsbibliothek 820, fols. 108r–115v; Paris, Bibliothèque de l'Arsenal 522, fols. 106r–109v; Paris, Bibliothèque Nationle, Latin Ms. 14580, fols. 82vb–86ra.

———. *Dicta*. Ms. Munich, Bayrische Staatsbibliothek, CLM 4721, fols. 199ra–202ra.

———. *Egregia puncta et notata de anima*. Ms. Erfurt, Wissenschaftliche Bibliothek der Stadt, Amplonianus F. 339, fols. 73ra–108ra.

———. *Epistola . . . ad episcopum Wormaciensem*. Ed. Gustav Sommerfeldt, "Die Prophetien der hl. *Hildegard von Bingen* in einem Schreiben des Magisters Heinrich v. Langenstein (1383), und Langensteins Trostbrief über den Tod eines Bruders des Wormser Bishofs Eckard von Ders (um 1384). *Historisches Jahrbuch* 30 (1909): 43–61, 297–307.

———. *Expositio terminorum astronomiae*. Mss. London, British Museum, Harley 941, fols. 51r–58r; Munich, Universität Bibliothek Q. 738, fols. 93r–95v.

———. *Postilla super Isaiam*. Ms. Erfurt, Wissenschaftliche Bibliothek der Stadt, Amplonianus F. 173, fols. 20r–95r.

———. *Quaestio de cometa*. Ed. Hubert Pruckner, *Studien*, pp. 89–138.

———. *Quaestiones*, misc. Mss. Paris, Bibliothèque Nationale, Latin Ms. 14580, fols. 86ra–100vb, Latin Ms. 16401, fols. 47r–54v, Latin Ms. 17496, fols. 81v–86r; Vatican, Biblioteca Apostolica Vaticana, Vat. Lat. 9369, fols. 26r–40r.

———. *Quaestiones quarti Sententiarum*. Mss. Alençon, Bibliothèque de Ville 144, fols. 1ra–140va; Vienna, Österreichische Nationalbibliothek, CVP 4319, fols. 145r–237v.

———. *Quaestiones super perspectivam*. Print. Valencia, 1503. Mss. Erfurt, Wissenschaftliche Bibliothek der Stadt, Amplonianus F. 380, fols.

29r–40v; Florence, Biblioteca Nazionale Centrale, Fondo Conventi Soppressi J.X.19, fols. 56r–85v; Paris, Bibliothèque de l'Arsenal 522, fols. 66r–87r; Vienna, Österreichische Nationalbibliothek, CVP 4992, fols. 169v–172v (fragment), CVP 5437, fols. 105r–160v.

_____. *Sermo de Sancta Katharina Virgine.* Edited by Albert Lang, *Divus Thomas* 26 (1948): 132–159.

_____. *Tractatus contra astrologos conjunctionistas de eventibus futurorum.* Edited by Hubert Pruckner, *Studien,* pp. 139–206.

_____. *Tractatus Venerabilis Magistri Hainrici de Hassia contra quendam Eremitam . . . nomine Theolophorum.* Edited by Bernard Pez, *Thesaurus anecdotorum novissimus.* Augsbourg, 1721, 2: 507–564.

_____. *Tractatus de discretione spirituum.* Vienna, Österreichische Nationalbibliothek, CVP 5086, fols. 99r–108v. This work is extant in over sixty manuscript copies.

_____. *Tractatus de horis canonicis.* Mss. Erfurt, Wissenschaftliche Bibliothek der Stadt, Amplonianus Q. 145, fols. 172r–179v, Amplonianus Q. 150, fols. 202r–204r; Munich, Bayrische Staatsbibliothek, CLM 5338, fols. 199r–206v, CLM 15173, fols. 228r–231v, CLM 15602, fols. 13r–21v; Cologne, Stadtbibliothek GB4.154, fols. 17r–27v.

_____. *Tractatus de sphaera.* Vatican, Biblioteca Apostolica Vaticana, Vat. Lat. 9369, fols. 41r–50r.

3. Other primary sources

d'Ailly, Pierre. *De impressionibus aeris.* Strassburg: Johann Prüss, 1504.

Albert the Great. *De anima.* Edited by Clemens Stroick, *Opera omnia.* Vol. 7, pt. 1. Monasterii Westfalorum in aedibus: Aschendorff, 1968.

_____. *De animalibus libri XXVI.* Edited by Hermann Stadler. Beiträge zur Geschichte der Philosophie des Mittelalters 15–16. Munich: Aschendorff, 1916, 1921.

_____. *De meteoris libri IV.* Edited by Augustine Borgnet, Vol. 4: 477–808, of *Opera omnia.* 38 Vols. Paris: Vivès, 1890–1899.

_____. *De natura locorum.* Edited by Borgnet. Vol. 9: 527–582.

_____. *Summae de creaturis.* Edited by Borgnet. Vols. 34 and 35.

Alexander of Hales. *Summa theologica.* Edited Quaracchi. Florence: Ex typographia Collegii S. Bonaventurae, 1924–1930.

Ambrose, *Hexaemeron.* CSEL 32: 3–261.

Apianus, Peter. *Cosmographiae introductio.* Paris: Guillaume Cavellat, 1550.

Aquinas, Thomas. *In libros Aristotelis Meteorologicam [Expositio].* Edited by Raymundi Spiazzi. Rome: Marietti, 1952.

____. *Summa contra gentiles.* Translated English Dominican Fathers. 4 Vols. London: Burns, Oates, and Washbourne, 1923–1929.

____. *Summa theologiae.* Edited Blackfriars. 60 Vols. New York: McGraw-Hill, 1964–.

Aristotle. *Opera.* Edited by Immanuel Bekker. 5 Vols. Berlin: Academia Regia Borussica, 1831–1870.

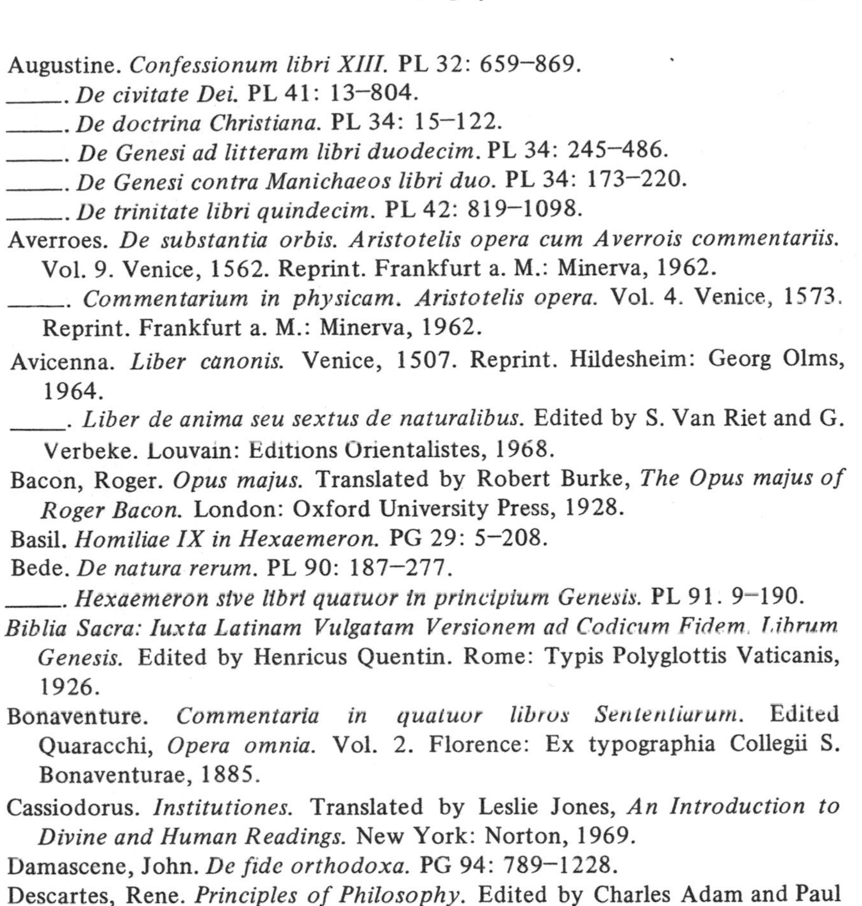

Augustine. *Confessionum libri XIII.* PL 32: 659–869.
_____. *De civitate Dei.* PL 41: 13–804.
_____. *De doctrina Christiana.* PL 34: 15–122.
_____. *De Genesi ad litteram libri duodecim.* PL 34: 245–486.
_____. *De Genesi contra Manichaeos libri duo.* PL 34: 173–220.
_____. *De trinitate libri quindecim.* PL 42: 819–1098.

Averroes. *De substantia orbis. Aristotelis opera cum Averrois commentariis.* Vol. 9. Venice, 1562. Reprint. Frankfurt a. M.: Minerva, 1962.

_____. *Commentarium in physicam. Aristotelis opera.* Vol. 4. Venice, 1573. Reprint. Frankfurt a. M.: Minerva, 1962.

Avicenna. *Liber canonis.* Venice, 1507. Reprint. Hildesheim: Georg Olms, 1964.

_____. *Liber de anima seu sextus de naturalibus.* Edited by S. Van Riet and G. Verbeke. Louvain: Editions Orientalistes, 1968.

Bacon, Roger. *Opus majus.* Translated by Robert Burke, *The Opus majus of Roger Bacon.* London: Oxford University Press, 1928.

Basil. *Homiliae IX in Hexaemeron.* PG 29: 5–208.

Bede. *De natura rerum.* PL 90: 187–277.

_____. *Hexaemeron sive libri quatuor in principium Genesis.* PL 91. 9–190.

Biblia Sacra: Iuxta Latinam Vulgatam Versionem ad Codicum Fidem, Librum Genesis. Edited by Henricus Quentin. Rome: Typis Polyglottis Vaticanis, 1926.

Bonaventure. *Commentaria in quatuor libros Sententiarum.* Edited Quaracchi, *Opera omnia.* Vol. 2. Florence: Ex typographia Collegii S. Bonaventurae, 1885.

Cassiodorus. *Institutiones.* Translated by Leslie Jones, *An Introduction to Divine and Human Readings.* New York: Norton, 1969.

Damascene, John. *De fide orthodoxa.* PG 94: 789–1228.

Descartes, Rene. *Principles of Philosophy.* Edited by Charles Adam and Paul Tannery, *Oeuvres de Descartes.* Vol. 9, pt. 2. Paris: Librairie philosophique J. Vrin, 1964.

Dionysius the Pseudo-Areopagite. *De coelesti hierarchia.* PG 3: 119–370.

_____. *De divinis nominibus.* PG 3: 586–996.

Durand of Saint-Pourçain. *Petri Lombardi Sententias theologicas commentariorum libri IIII.* Venice, 1571. Reprint. Ridgewood, N.J.: Gregg Press, 1964.

Facinus de Ast. *Sententiae.* Ms. Erfurt, Wissenschaftliche Bibliothek der Stadt, Amplonianus F. 115.

Geber (Jabir ibn Aflaḥ). *De astronomia libri IX.* Nuremberg: Johann Petrejus, 1534.

Gilbert, William. *De magnete magneticisque corporibus et de magno magnete tellure physiologia nova.* Translated by P. Mottelay, *Great Books of the Western World.* Vol. 28. Chicago: William Benton, 1952.

Hugh of St. Victor. *Didascalicon.* Translated by Jerome Taylor, *Didascalicon; a Medieval Guide to the Arts.* New York: Columbia University Press, 1961.

_____. *De sacramentis Christianae fidei.* PL 176: 173–618.

Isidore of Seville. *Etymologiarum sive originum libri XX.* Edited by W.M. Lindsay. 2 Vols. Oxford: Clarendon Press, 1911 (contains no pagination).

Lombard, Peter. *Sententiae.* Text in Bonaventure, *Commentaria in quatuor libros Sententiarum.* Florence, 1885.

Lucretius. *De rerum natura.* Translated by William J. Callaghan, *Lucretius on the Nature of the Universe.* Boston, Mass.: Student Outline, 1964.

Luther, Martin. *Lectures on Genesis.* Edited by Jaroslav Pelikan, *Luther's Works.* Saint Louis, Mo.: Concordia Publishing House, 1958.

Macrobius. *Commentarii in Somnium Scipionis.* Edited by Ludovicus Ianus, *Macrobii Ambrosii Theodosii . . . Opera.* Vol. 1. Leipzig: Godofredus Bassius, 1898.

Nicholas of Lyra. *Postilla super totam Bibliam.* Nuremberg: Anthon Koberger, 1494.

Nicholas of Autrecourt. *Exigit ordo.* Translated by Leonard Kennedy et al., *The Universal Treatise of Nicholas of Autrecourt.* Milwaukee, Wis.: Marquette University Press, 1971.

Oresme, Nicole, *Le Livre du ciel et du monde.* Edited by Albert Menut and Alexander Denomy. Madison: University of Wisconsin Press, 1968.

_____. *Tractatus de configurationibus qualitatum et motuum.* Edited by Marshall Clagett, *Nicole Oresme.* See secondary bibliography.

Pecham, John. *Perspectiva communis.* Edited by David Lindberg, *John Pecham and the Science of Optics.* Madison, Wis.: University of Wisconsin Press, 1970.

_____. *Tractatus de perspectiva.* Edited by David Lindberg, Franciscan Institute Publications, Text Series no. 16. St. Bonaventure, N.Y.: Franciscan Institute, 1972.

Petrarch, Francesco. *De ignorantia.* Translated by Hans Nachod, *The Renaissance Philosophy of Man.* Chicago: University of Chicago Press, 1948.

Plato. *Phaedo.* Translated by B. Jowett, *The Dialogues of Plato.* Vol. 1. Oxford: Clarendon, 1871.

_____. *Timaeus.* Translated by Francis Cornford, *Plato's Cosmology.* New York: Harcourt, Brace, 1937.

Pliny. *Naturalis historiae.* Edited by H. Rackham et al. Loeb Classical Library. 10 Vols. Cambridge, Mass.: Harvard University Press, 1938–1962.

Ptolemy. *Almagestum.* Translated by R.C. Taliaferro, *Great Books of the Western World.* Vol. 16. Chicago: William Benton, 1952.

Sacrobosco. *De sphaera.* Translated by Lynn Thorndike, *The Sphaera of Sacrobosco and Its Commentators.* Chicago: University of Chicago Press, 1949.

_____. *Sphaera Ioannis de Sacrobosco, emendata Eliae Vineti Santonis.* Venice: Herede Hieronymi Scoli, 1536.

Strabo, Walafrid. *Glossa ordinaria, Liber Genesis.* PL 113: 67–182.

Thomas of Strassbourg. *Commentaria in IIII libros Sententiarum.* Venice, 1564. Reprint. Ridgewood, N.J.: Gregg Press, 1965.

Vincent of Beauvais. *Speculum naturale.* [Strassburg: Printer of the 'Legenda aurea', about 1481].

William of Ockham. *Super 4 libros Sententiarum.* In *Opera plurima.* Vols. 3–4. Lyon, 1494–1496. Reprint. London: Gregg Press, 1962.

Secondary Sources

Alessio, Franco. "Causalita' naturale e causalita' divina nel 'De habitudine causarum' di Enrico de Langenstein." In *La Filosofia della natura nel medioevo,* Atti del Terzo Congresso Internazionale di Filosofia Medioevale. Milan: Società editrice Vita e Pensiero, 1966.

Apfaltrer, Ernst. *Scriptores antiquissimae ac celeberrimae Universitatis Viennensis.* Vol. 1. Vienna, 1740.

Baur, Ludwig. *Die philosophischen Werke des Robert Grosseteste, Bischofs von Lincoln.* Beiträge zur Geschichte der Philosophie des Mittelalters 9. Munich: Aschendorff, 1912.

Beazley, Charles R. *The Dawn of Modern Geography: A History of Exploration and Geographical Science.* 3 Vols. London: J. Murray, 1897–1906.

Benjamin, Francis S. and Toomer, G.J. *Campanus of Novara and Medieval Planetary Theory: Theorica planetarum.* Madison, Wis.: University of Wisconsin Press, 1971.

Berger, Samuel. "Les préfaces jointes aux livres de la Bible dans les manuscrits de la Vulgate," *Memoires présentes par divers savants à l'Academie des Inscriptions et Belles–lettres* 1st ser. 11, pt. 2 (Paris, 1904): 1–78.

Berthelot, Marcelin. *Histoire des sciences. La Chimie au moyen âge.* Paris: Imprimerie Nationale, 1893.

Brampton, Charles. "Scotus, Ockham and the Theory of Intuitive Cognition." *Antonianum* 40 (1965): 449–466.

Boyer, Carl. *The Rainbow: From Myth to Mathematics.* New York: Thomas Yoseloff, 1959.

Brett, George. *A History of Psychology.* 3 Vols. London: G. Allen, 1912–1921.

Burke, Peter. *The Renaissance Sense of the Past.* London: Edward Arnold, 1969.

Burtt, Edwin A. *The Metaphysical Foundations of Modern Science.* Revised ed. Garden City, N.Y.: Doubleday, 1954.

Cheyney, Edward. *The Dawn of a New Era, 1250–1453.* New York: Harper Brothers, 1936; Torchbook ed., 1962.

Clagett, Marshall. *Nicole Oresme and the Medieval Geometry of Qualities of Motions.* Madison, Wis.: University of Wisconsin Press, 1968.

_____. "Richard Swineshead and Late Medieval Physics." *Osiris* 9 (1950): 131–140.

_____. *The Science of Mechanics in the Middle Ages.* Madison, Wis.: University of Wisconsin Press, 1959.

———. "Some General Aspects of Medieval Physics." *Isis* 39 (1948): 29–44.

Cohen, Morris and Drabkin, I.E., eds. *A Source Book in Greek Science.* Cambridge, Mass.: Harvard University Press, 1958.

Collingwood, Robin G. *The Idea of Nature.* New York: Galaxy Books, 1960.

Collison, Robert. *Encyclopaedias: Their History Throughout the Ages.* New York: Hafner, 1964.

Coopland, George W. *Nicole Oresme and the Astrologers.* Liverpool: University Press of Liverpool, 1952.

Copleston, Frederick. *A History of Philosophy.* Vol. 2, part 1. New York: Image Books, 1962.

Courtenay, William. "Covenant and Causality in Pierre d'Ailly." *Speculum* 46 (1971): 94–119.

———. "Nominalism and Late Medieval Religion." In *The Pursuit of Holiness in Late Medieval and Renaissance Religion,* edited by Charles Trinkaus with Heiko Oberman. Leyden: Brill, 1974.

———. "Nominalism and Late Medieval Thought: A Bibliographical Essay." *Theological Studies* 33 (1972): 716–734.

Crombie, Alastair. *Medieval and Early Modern Science.* 2nd ed. revised. 2 Vols. Garden City, N.Y.: Doubleday, 1959.

———. *Robert Grosseteste and the Origins of Experimental Science,* 1100–1700. Oxford: Clarendon Press, 1953.

Day, Sebastian. *Intuitive Cognition: A Key to the Significance of the Later Scholastics.* Franciscan Institute Publications, Philosophy Series no. 4. St. Bonaventure, N.Y.: Franciscan Institute, 1947.

Delisle, Léopold. *Recherches sur la librarie de Charles V.* 2 Vols. Paris: H. Champion, 1907.

Denifle, Heinrich and Aemilio Chatelain, eds. *Auctarium chartularii Universitatis Parisiensis.* Vol. 1. Paris: Fratres Delalain, 1894.

———. *Chartularium Universitatis Parisiensis.* Vols. 1 and 2. Paris: Fratres Delalain, 1889, 1891.

Donahue, William H. "The Solid Planetary Spheres in Post-Copernican Natural Philosophy." In *The Copernican Achievement,* edited by Robert S. Westman, forthcoming.

Dougherty, Kenneth. *Cosmology: An Introduction to the Thomistic Philosophy of Nature.* Peekskill, N.Y.: Graymoor Press, 1953.

Dreyer, John. *A History of Astronomy from Thales to Kepler.* 2nd ed. New York: Dover, 1953.

Duhem, Pierre. *Études sur Léonard de Vinci.* 3 Vols. Paris: Hermann, 1906–1913.

———. *To Save the Phenomena.* Translated by E. Doland and C. Maschler. Chicago: University of Chicago Press, 1969.

———. *Le Système du monde.* 10 Vols. Vols. 1–5, Paris: Hermann, 1913–1916; Vols. 6–10, Paris: Hermann, 1954–1959.

Eliade, Mircea. *The Myth of the Eternal Return.* Translated by Willard R. Trask. New York: Pantheon Books, 1954.

Eslick, Leonard. "The Material Substrate in Plato." In *The Concept of Matter in Greek and Medieval Philosophy,* edited by Ernan McMullin. Notre Dame, Ind.: University of Notre Dame Press, 1965.

Flint, Robert. *Philosophy as Scientia Scientiarum, and A History of Classifications of the Sciences.* London: William Blackwood, 1904.

Foucault, Michel. *L'Archéologie du savior.* Paris: Gallimard, 1969.

_____. *Les mots et les choses.* Paris: Gallimard, 1966.

Frankfort, Henri et al. *Before Philosophy: The Intellectual Adventure of Ancient Man.* Baltimore, Md.: Penguin Books, 1966.

Funkenstein, Amos. "The Dialectical Preparation for Scientific Revolutions." In *The Copernican Achievement,* edited by Robert S. Westman, forthcoming.

Garin, Eugenio. *Science and Civic Life in the Italian Renaissance.* Translated by Peter Munz. Garden City, N.Y.: Anchor Books, 1969.

Gillispie, Charles C. *The Edge of Objectivity.* Princeton: Princeton University Press, 1960.

Glorieux, Palémon. "Sentences (Commentaires sur les)." *Dictionnaire de théologie catholique* 14 (Paris, 1941): 1875–1884.

Grant, Edward. "Hypotheses in Late Medieval and Early Modern Science." *Daedalus* 91 (1962): 599–616.

_____. "Late Medieval Thought, Copernicus and the Scientific Revolution." *Journal of the History of Ideas* 23 (1962): 197–220.

_____. *Physical Science in the Middle Ages.* New York: John Wiley, 1971.

_____, ed. *A Source Book in Medieval Science.* Cambridge, Mass.: Harvard University Press, 1974.

Hall, Marie (Boas). *The Scientific Renaissance, 1450–1630.* New York: Harper, 1962.

Hartwig, Otto. *Henricus de Langenstein dictus de Hassia: Zwei Untersuchungen über das Leben und die Schriften Heinrichs von Langenstein.* Marburg: Elwert'sche Universitäts–Buchhandlung, 1857.

Heilig, Konrad. "Heinrich Heimbuche von Langenstein." *Lexikon für Theologie und Kirche* 4 (1932): 924–925.

_____. "Kritische Studien zum Schrifttum der beiden Heinriche von Hessen." *Römische Quartalschrift* 14 (1943): 105–176.

Holton, Gerald. *Thematic Origins of Scientific Thought.* Cambridge, Mass.: Harvard University Press, 1973.

Jaki, Stanley L. *Science and Creation: From Eternal Cycles to an Oscillating Universe.* New York: Science History Publications, 1974.

Kargon, Robert H. *Atomism in England from Hariot to Newton.* Oxford: Clarendon Press, 1966.

Kearney, Hugh. *Science and Change, 1500–1700.* New York: McGraw–Hill, 1971.

Kimble, George H.T. *Geography in the Middle Ages.* London: Methuen, 1938. Reprint. New York: Russell and Russell, 1968.

Klubertanz, George. *The Discursive Power: Sources and Doctrine of the 'Vis Cogitativa' According to St. Thomas Aquinas.* St. Louis: Modern Schoolman, 1952.

Koyré, Alexandre. *Newtonian Studies.* Cambridge, Mass.: Harvard University Press, 1965.

Kren, Claudia. "Homocentric Astronomy in the Latin West. The *De reprobatione ecentricorum et epiciclorum* of Henry Hesse." *Isis* 59 (1968): 269–281.

_____. "A Medieval Objection to 'Ptolemy'." *British Journal for the History of Science* 4 (1969): 378–393.

_____. "The *Questiones super de celo* of Nicole Oresme." Ph.D. dissertation, The University of Wisconsin, Madison, 1965.

Kuhn, Thomas. *The Copernican Revolution.* New York: Vintage, 1957.

Lang, Albert. "Die Katharinenpredigt Heinrichs von Langenstein." *Divus Thomas* 26 (1948): 123–159, 233–258, 361–394.

_____. *Die Wege der Glaubensbegründung bei den Scholastikern des 14. Jahrhunderts.* Beiträge zur Geschichte der Philosophie des Mittelalters 30, nos. 1,2. Munich: Aschendorff, 1930.

Lang, Justin. *Die Christologie bei Heinrich von Langenstein: Eine dogmenhistorische Untersuchung.* Freiburger theologische Studien 85. Freiburg: Herder, 1966.

Lappe, Josef. *Nicolaus von Autrecourt: Sein Leben, seine Philosophie, seine Schriften.* Beiträge zur Geschichte der Philosophie des Mittelalters 6, no. 2. Munich: Aschendorff, 1908.

Lasswitz, Kurd. *Geschichte der Atomistik vom Mittelalter bis Newton.* 2 Vols. Leipzig: Leopold Voss, 1890.

Lewis, C.S. *The Discarded Image: An Introduction to Medieval and Renaissance Literature.* Cambridge: University Press, 1964.

Lindberg, David C. "Alhazen's Theory of Vision and Its Reception in the West." *Isis* 58 (1967): 321–341.

_____. *A Catalogue of Medieval and Renaissance Optical Manuscripts.* Toronto: Pontifical Institute, 1975.

_____. *Theories of Vision from Alkindi to Kepler.* Forthcoming.

_____. "The Theory of Pinhole Images in the Fourteenth Century." *Archive for History of Exact Science* 6 (1970): 299–325.

_____ with Nicholas H. Steneck. "The Sense of Vision and the Origins of Modern Science." In *Science, Medicine and Society in the Renaissance,* edited by Allen G. Debus. 2 Vols. New York: Neale Watson, 1972.

Lohr, Charles. "Medieval Latin Aristotle Commentaries." *Traditio* 23 (1967): 313–413, and continuing.

Lovejoy, Arthur. *The Great Chain of Being.* Cambridge, Mass.: Harvard University Press, 1936. Reprint. New York: Harper, 1960.

Mabilleau, Léopold. *Histoire de la philosophie atomistique.* Paris: Imprimerie Nationale, 1895.

Maier, Anneliese. "Das Problem der 'species sensibiles in medio' und die neue Naturphilosophie des 14. Jahrhunderts." In *Ausgehendes Mittelalter.* 2 Vols. Rome: Edizioni di Storia e Letteratura, 1964–1967.

_____. *Zwei Grundprobleme der scholastischen Naturphilosophie.* 2nd ed. Rome: Edizioni di Storia e Letteratura, 1951.

Mangenot, E. "Genèse." *Dictionnaire de théologie catholique* 6 (Paris, 1915): 1206–1208.

McColley, Grant. "The Theory of the Diurnal Rotation of the Earth." *Isis* 26 (1936–1937): 392–402.

Moody, E.A. "Galileo and Avempace: The Dynamics of the Leaning Tower Experiment." *Journal of the History of Ideas* 12 (1951): 163–193, 375–422.

McKeough, Michael. *The Meaning of the rationes seminales in St. Augustine.* Catholic University of America, Philosophical Studies 15. Washington, D.C.: Catholic University, 1926.

O'Donnell, J.R. "The Philosophy of Nicolaus of Autrecourt and His Appraisal of Aristotle." *Mediaeval Studies* 4 (1942): 97–125.

Ovenden, Michael. "Intimations of Unity." In *Science and Society: Past, Present, and Future,* edited by Nicholas H. Steneck. Ann Arbor, Mi.: University of Michigan Press, 1975.

Paetow, Louis. *The Arts Course at Medieval Universities with Special Reference to Grammar and Rhetoric.* University of Illinois Studies 3, no. 7. Urbana, Ill.: University Press, 1910.

Panofsky, Erwin. *The Codex Huygens and Leonardo da Vinci's Art Theory.* London: The Warburg Institute, 1940. Reprint. Westport, Conn.: Greenwood Press, 1971.

_____. *Gothic Architecture and Scholasticism.* New York: Meridian Books, 1957.

Partington, James R. *A History of Chemistry.* Vol. 1. London: Macmillan, 1961.

Pedersen, Olaf. "Theorica: A Study in Language and Civilization." *Classica et Mediaevalia* 22 (1961): 151–166.

_____. "The Theorica Planetarum Literature of the Middle Ages." *Classica et Mediaevalia* 23 (1962): 225–232.

Pirzio, Paola. "Le prospettive filosofiche del trattato di Enrico de Langenstein (1325–1397) 'De habitude causarum'." *Revista critica di storia della filosofia* 24 (1969): 363–373.

Pruckner, Hubert. *Studien zu den astrologischen Schriften des Heinrich von Langenstein.* Leipzig: B.G. Teubner, 1933.

Randall, John Herman, Jr. *The School of Padua and the Emergence of Modern Science.* Padua: Editrice Antenore, 1961.

Rashdall, Hastings. *The Universities of Europe in the Middle Ages.* 3 Vols. Newly edited by F.M. Powicke and A.B. Emden. Oxford: Clarendon Press, 1936.

de Raedemaeker, Jozef. "Une ébauche de catalogue des commentaries sur le 'De anima' parus aux XIIIe, XIVe, et XVe siècles." *Bulletin de philosophie médiévale* 5 (1963): 149–183; 6 (1964): 119–134.

Robbins, Frank. *The Hexaemeral Literature: A Study of the Greek and Latin Commentaries on Genesis.* Chicago: University of Chicago Press, 1912.

Roth, F.W.E. "Zur Bibliographie des H. von Heimbuche de Hassia, dictus

Langenstein." *Beiheft zum Zentralblatt für Bibliothekswissenschaft* 1 (1888–1889): 97–118.

Santarem, Manuel. *Essai sur l'histoire de la cosmographie et de la cartographie pendant le Moyen-Age, et sur les progrès de la géographie après les grandes découvertes du XVe siècle.* 3 Vols. Paris: Maulde et Renou, 1849–1852.

de Santillana, Giorgio. *The Origins of Scientific Thought.* New York: Mentor, 1961.

Scott, Theodore. "John Buridan on the Objects of Demonstrative Science." *Speculum* 40 (1965): 654–673.

Singer, Charles. *Greek Biology and Greek Medicine.* Oxford: Clarendon Press, 1922.

Smalley, Beryl. *Historians in the Middle Ages.* London: Thames and Hudson, 1974.

_____. *The Study of the Bible in the Middle Ages.* Notre Dame: University of Notre Dame Press, 1964.

Stegmüller, Friedrich. *Repertorium Biblicum Medii Aevi.* 7 Vols. Madrid: Grafica Marina, 1951.

Steneck, Nicholas H. "Albert the Great on the Classification and Localization of the Internal Senses." *Isis* 65 (1974): 193–211.

_____. "A Late Medieval *Arbor scientiarum.*" *Speculum* 50 (1975): 245–269.

_____. "A Late Medieval Debate concerning the Primary Organ of Perception." *Proceedings of the XIIIth International Congress of the History of Science* 3, 4 (Moscow, 1974): 198–204.

_____. "The Problem of the Internal Senses in the Fourteenth Century." Ph.D. dissertation. The University of Wisconsin, 1970.

Stock, Brian. *Myth and Science in the Twelfth Century: A Study of Bernard Silvester.* Princeton: Princeton University Press, 1972.

Thayer, H.S. *Newton's Philosophy of Nature.* New York: Hafner, 1965.

Thorndike, Lynn. *A History of Magic and Experimental Science.* Vol. 3. New York: Columbia University Press, 1934.

Trapp, Damascus. "Augustinian Theology of the 14th Century." *Augustiniana* 6 (1956): 146–274.

Trinkaus, Charles. *In Our Image and Likeness.* 2 Vols. Chicago: University of Chicago Press, 1970.

Vanderbroucke, François. "Henri de Langenstein." *Dictionnaire de spiritualite* 7 (Paris, 1969): 205–219.

Van Melsen, Andrew. *From Atomos to Atom.* New York: Harper, 1960.

Wallace, William A. *Causality and Scientific Explanation.* 2 Vols. Ann Arbor: University of Michigan Press, 1972–1974.

_____. "The Enigma of Domingo de Soto: *Uniformiter difformis* and Falling Bodies in Late Medieval Physics." *Isis* 59 (1968): 384–401.

_____. *St. Thomas Aquinas, Summa Theologiae:* Vol. 10 *Cosmogony.* New York: McGraw-Hill, 1967.

Weinberg, Julius. *Nicolaus of Autrecourt: A Study in 14th Century Thought.* Princeton: Princeton University Press, 1948.

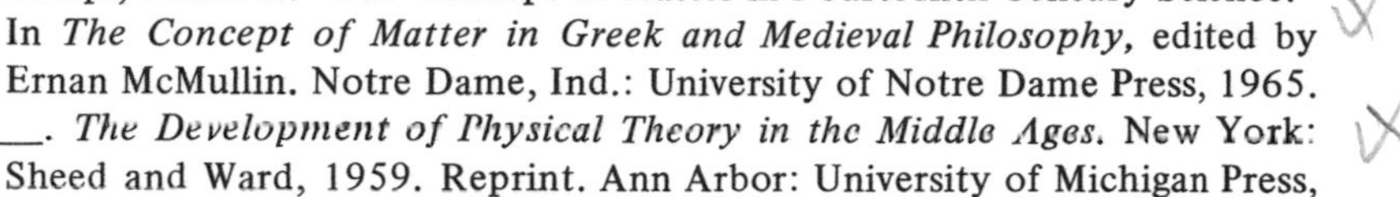

Weisheipl, James A. "The Concept of Matter in Fourteenth Century Science." In *The Concept of Matter in Greek and Medieval Philosophy,* edited by Ernan McMullin. Notre Dame, Ind.: University of Notre Dame Press, 1965.

____. *The Development of Physical Theory in the Middle Ages.* New York: Sheed and Ward, 1959. Reprint. Ann Arbor: University of Michigan Press, 1971.

____. "Developments in the Arts Curriculum at Oxford in the Early Fourteenth Century." *Mediaeval Studies* 28 (1966): 151–175.

____. "Ockham and Some Mertonians." *Mediaeval Studies* 30 (1968): 163–213.

Wolfson, Harry. "The Internal Senses in Latin, Arabic, and Hebrew Philosophical Texts." *Harvard Theological Review* 28 (1935): 69–133.

Index